Ecology and the World-System

Studies in the Political Economy of the World-System (Formerly published as Political Economy of the World-System Annuals)

Rethinking the Nineteenth Century: Contradictions and Movements
edited by Francisco O. Ramirez

Racism, Sexism, and the World-System
edited by Joan Smith, Jane Collins, Terence K. Hopkins, and Akbar Muhammad

Revolution in the World-System
edited by Terry Boswell

War in the World-System
edited by Robert K. Schaeffer

Semiperipheral States in the World-Economy
edited by William G. Martin

Cities in the World-System
edited by Resat Kasaba

Pacific-Asia and the Future of the World-System
edited by Ravi Arvind Palat

Commodity Chains and Global Capitalism
edited by Gary Gereffi and Miguel Korzeniewicz

Food and Agrarian Orders in the World-Economy
edited by Philip McMichael

A New World Order?: Global Transformations in the Late Twentieth Century
edited by David A. Smith and József Böröcz

Latin America in the World-Economy
edited by Roberto Patricio Korzeniewicz and William C. Smith

Space and Transport in the World-System
edited by Paul S. Ciccantell and Stephen G. Bunker

Ecology and the World-System

edited by

Walter L. Goldfrank, David Goodman,
and Andrew Szasz

Contributions in Economics and Economic History, Number 211

Studies in the Political Economy of the World-System
Immanuel Wallerstein, Series Adviser

GREENWOOD PRESS
Westport, Connecticut • London

Library of Congress Cataloging-in-Publication Data

Ecology and the world-system / edited by Walter L. Goldfrank, David Goodman, and Andrew Szasz.
p. cm. — (Contributions in economics and economic history, ISSN 0084-9235 ; 211) (Studies in the political economy of the world-system)
Includes bibliographical references and index.
ISBN 0-313-30725-3 (alk. paper)
1. Environmental economics. 2. Biosphere. I. Goldfrank, Walter L. II. Goodman, David, 1938– . III. Szasz, Andrew, 1947– IV. Series. V. Series: Studies in the political economy of the world-system.
HC79.E5 E2165 1999
333.7—dc21 98-47131

British Library Cataloguing in Publication Data is available.

Library of Congress Catalog Card Number: 98-47131
ISBN: 0-313-30725-3
ISSN: 0084-9235

First published in 1999

Greenwood Press, 88 Post Road West, Westport, CT 06881
An imprint of Greenwood Publishing Group, Inc.
www.greenwood.com

Printed in the United States of America

The paper used in this book complies with the Permanent Paper Standard issued by the National Information Standards Organization (Z39.48–1984).

10 9 8 7 6 5 4 3 2 1

Copyright Acknowledgments

Every reasonable effort has been made to trace the owners of copyright materials in this book, but in some instances this has proven impossible. The author and publisher will be glad to receive information leading to more complete acknowledgments in subsequent printings of the book and in the meantime extend their apologies for any omissions.

Contents

Illustrations

Figures

Tables

Maps

Series Foreword

Immanuel Wallerstein

The Political Economy of the World-System Section of the American Sociological Association was created in the 1970s to bring together a small but growing number of social scientists concerned with analyzing the processes of world-systems in general, and our modern one in particular.

Although organizationally located within the American Sociological Association, the PEWS Section bases its work on the relative insignificance of the traditional disciplinary boundaries. For that reason it has held an annual spring conference, open to and drawing participation from persons who work under multiple disciplinary labels.

For PEWS members, not only is our work unidisciplinary, but the study of the world-system is not simply another "specialty" to be placed beside so many others. It is instead a different "perspective" with which to analyze all the traditional issues of the social sciences. Hence, the themes of successive PEWS conferences are quite varied and cover a wide gamut of topics. What they share is the sense that the isolation of political, economic, and sociocultural "variables" is a dubious enterprise, that all analysis must be simultaneously historical and systemic, and that the conceptual bases of work in the historical social sciences must be rethought.

Acknowledgments

Many people work behind the scenes to make a book possible. In this instance, the editors would like to thank the faculty, staff, and students at the University of California, Santa Cruz who provided financial and logistical support for the twenty-first annual conference of the Political Economy of the World-System (PEWS) Section of the American Sociological Association. The conference, from which the chapters that follow were selected, took place April 3–5, 1997 and was subsidized by the Division of Social Sciences and the Center for Cultural Studies. We were particularly fortunate to have the irrepressible and omnicompetent Karyn Amy Riegel serve as conference administrator. Cheryl Van DeVeer and Zoe Sodja of the Document Publishing and Editing Center prepared the final manuscript.

Introduction

In organizing this volume, we have sought to emphasize three ways in which environmental analysis intersects with the long-standing concerns of scholars working within the world-systems framework, if not always explicitly so. These are (1) the emergence of threats to the global environment and of ecological limits to the sustainability of capitalism; (2) the differences in environmental impacts among different types of historical systems and among different parts of the capitalist world-economy; and (3) replication and variation among environmental social movements in the contemporary world. At the same time, we intend that this volume demonstrate ways in which the consideration of environmental issues can enrich accounts of change at the local, regional, national, and world levels, by showing how ecological threats and limits rebound upon the systems that generate them, thus undermining their capacity to reproduce themselves.

Prior to the second half of our century, environmental damage typically occurred on a local or regional level. But with its unprecedented impacts on the biosphere, the most recent phase of world-system development has introduced new concerns: global warming, also known as the greenhouse effect; ozone depletion; altered oceanic ecosystems. Along with the danger of widespread nuclear war, these global trends introduce the real possibility that the continuation of human life itself may be measured in decades or centuries rather than millennia.

In less dramatic ways, but no less inexorably, the accelerating commodification of nature makes livelihood problematic for large numbers of people in many parts of the world, through desertification and salinization of land; depletion and pollution of water; destruction of wetlands and fisheries; and harvesting or burning of forests. Although the measurement of environmental damage is an enterprise fraught with controversy, there are virtually no reputable analysts who claim that the world as a whole as well as many parts of it are not deteriorating environmentally. Rather, the questions are how much and how fast they are deteriorating, how imminent are which dangers, and with what consequences for the system and its parts.

The chapters in Part I address issues at a global level. Immanuel Wallerstein's chapter argues that capitalism as an historical system is reaching its limits largely because it is simultaneously exhausting its reserves of nonpolluted, nondepleted land and nonproletarianized labor. The more labor costs rise from increasing deruralization, the greater the pressure on capitalists to continue externalizing other costs by neglecting and delaying environmental restoration and protection. Wallerstein's prediction of doom (at least for capitalism) is both modified and specified in Peter Grimes's essay. Grimes offers two scenarios, neither of them appetizing: either continued use of fossil fuels overheats the atmosphere and in effect cooks us; or scarcity of fossil fuels (some uses of which cannot be replaced by alternative "soft" energy sources) engenders increasingly severe conflicts.

Although it shares their spirit, Albert J. Bergesen and Laura Parisi's chapter goes in two directions that differ significantly from the general or specific predictions of inevitably terminal capitalist crisis. On the one hand, they report and interpret a number of empirical findings correlating the level of toxic emissions with such time-honored independent variables as level of foreign investment and level of commodity concentration. Their findings suggest the problem that what is better for the environment is worse for "development," at least as that process is conventionally understood at the national level. These findings also connect back to the chapter's second direction, the argument that increased environmental justice for humans may be incompatible with improving what the authors call "justice for the environment."

In the last chapter in Part I, J. Timmons Roberts joins Grimes to present an overview and and agenda of important ways world-system concepts and approaches can enrich environmental analyses, to which they have not been much addressed until very recently. This review of both environmental and world-systems research documents a growing convergence. In addition, it proposes ways in which "subsidies" from the environment should be more prominently featured in world-system analyses themselves.

Taken together, these chapters suggest the frustrating paradox that our collective intellectual capacity to understand threats to the world environment is growing precisely at a time when our collective political capacity to counter them is distressingly small, thanks to the (at least) short-term triumph of neoliberal ideology and its carriers.

Part II exemplifies the vitality and variety of Roberts and Grimes's agenda by offering five chapters with a delimited scope in time and space. Sing Chew combines ecological and world-system analyses to explain the decline of the Bronze Age civilizations of Mesopotamia and Harappa (Incus Valley) after 1700 B.C. Stephen Bunker and Paul Ciccantell demonstrate the parallel ways in which ascendant hegemons over the 400-year course of the capitalist world-economy reorganized material relations with nature as an essential aspect of their movement toward the top. Ilmo Massa traces the changing environmental impacts of capitalist expansion on the circumpolar north of the planet and explicates the current triangular conflict among developers, conservationists, and native peoples. Gavan McCormack explores the deleterious social and ecological consequences of Asians' eager adoption of modernist "large dam" hydroelectric projects, first in Japan and more recently in China and Vietnam. Zsuzsa Gille, finally, analyzes the production and conceptualization of industrial waste in state socialist Hungary, as well as fundamental continuities in practice into the "post-socialist" period.

The chapters in Part III shift the focus to environmental movements in the United States and in two semiperpheral countries (South Africa and South Korea), exemplars of the zone to which much heavy industry has migrated in the last 25 years. Robert Schaeffer explains how and why the U.S. movement won a succession of policy victories in the 1970s only to find its progress stalled in the subsequent period. Christine Root and David Wiley analyze the limited successes of a grassroots anti-toxics movement in Durban, showing the acute contradictions between the post-apartheid government's commitments to its social base and its promotion of renewed economic growth. Su-Hoon Lee and David Smith portray South Korean environmentalism as moderately successful in changing public attitudes about the need to protect nature but virtually helpless in the face of underregulated industries that routinely produce ecological devastation.

From the civilizational past to the (possibly apocalyptic) future, from the biospheric and global to the nitty-gritty and local, these chapters represent the multiple and various ways contemporary scholars are grappling with questions of environmental and social change. We hope that they inspire fruitful inquiry and advance the pursuit of a sustainable world-system.

Part I

1

Ecology and Capitalist Costs of Production: No Exit

Immanuel Wallerstein

Today, virtually everyone agrees that there has been a serious degradation of the natural environment in which we live, by comparison with 30 years ago, *a fortiori* by comparison with 100 years ago, not to speak of 500 years ago. And this is the case, despite the fact that there have been continuous significant technological inventions and an expansion of scientific knowledge that one might have expected would have led to the opposite consequence. As a result, today, unlike 30 or 100 or 500 years ago, ecology has become a serious political issue in many parts of the world. There are even reasonably significant political movements organized centrally around the theme of defending the environment against further degradation and reversing the situation to the extent possible.

Of course, the appreciation of the degree of seriousness of the contemporary problem ranges from those who consider doomsday imminent to those who consider that the problem is one well within the possibility of an early technical solution. I believe the majority of persons hold a position somewhere in-between. I am in no position to argue the issue from a scientific viewpoint. I will take this in-between appreciation as plausible and will engage in an analysis of the relevance of this issue to the political economy of the world-system.

The entire process of the universe is, of course, one of unceasing change, so the mere fact that things are not what they were previously is so banal that it merits no notice whatsoever. Furthermore, within this constant turbulence, there are patterns of structural renewal we call life. Living, or organic, phenomena have a beginning and an end to their individual existence, but in the process procreate so that the species tends to continue. But this cyclical renewal is never perfect, and the overall ecology is therefore never static. In addition, all living phenomena ingest in some way products external to them, including, most of the time, other living phenomena, and predator/prey ratios are never perfect so that the biological milieu is constantly evolving.

Furthermore, poisons are natural phenomena as well and were playing a role in the ecological balance sheets long before human beings got into the picture. To be sure, today we know so much more chemistry and biology than our ancestors did that we are perhaps more conscious of the toxins in our environment; although perhaps not, since we are also learning these days how sophisticated the preliterate peoples were about toxins and antitoxins. We learn all these things in our primary and secondary school education and from the simple observation of everyday living. Yet often we tend to neglect these obvious constraints when we discuss the politics of ecological issues.

The only reason it is worth discussing these issues at all is if we believe that something special or additional has been happening in recent years, a level of increased danger, and if at the same time we believe that it is possible to do something about this increased danger. The case that is generally made by the green and other ecology movements precisely comprises both these arguments: increased level of danger (for example, holes in the ozone layer, greenhouse effects, or atomic meltdowns) and potential solutions.

As I said, I am willing to start on the assumption that there is a reasonable case for increased danger, one that requires some urgent reaction. However, in order to be intelligent about how to react to danger, we need to ask two questions: for whom does the danger exist? and what explains the increased danger? The "danger for whom" question has in turn two components: whom amongst human beings and whom amongst living beings. The first question raises the comparison of North-South attitudes on ecological questions; the second is the issue of deep ecology. Both in fact involve issues about the nature of capitalist civilization and the functioning of the capitalist world-economy, which means that before we can address the issue of "for whom," we had better analyze the source of the increased danger.

The story begins with two elementary features of historical capitalism. One is well-known: capitalism is a system that has an imperative need to expand—in terms of total production and geographically—in order to sustain its prime objective, the endless accumulation of capital. The second feature is less often discussed. An essential element in the accumulation of capital is for capitalists, especially large capitalists, not to pay their bills. This is what I call the "dirty secret" of capitalism.

Let me elaborate these two points. The first, the constant expansion of the capitalist world-economy, is admitted by everyone. The defenders of capitalism tout it as one of its great virtues. Persons concerned with ecological problems point to it as one of its great vices and in particular often discuss one of the ideological underpinnings of this expansion, which is the assertion of the right (indeed duty) of human beings "to conquer nature." Now, to be sure, neither expansion nor the conquest of nature was unknown before the onset of the capitalist world-economy in the sixteenth century. But, like many other things that were social phenomena prior to this time, neither had existential priority in previous historical systems. What historical capitalism did was to push these two themes—the actual expansion and its ideological justification—to the forefront, and thus capitalists were able to override social objections to this terrible duo. This is the real difference between historical capitalism and previous historical systems. All the values of capitalist civilization are millennial, but so are other contradictory values. What we mean by historical capitalism is a system in which the institutions that were constructed made it possible for capitalist values to take priority, such that the world-economy was set upon the path of the commodification of everything in order that there be ceaseless accumulation of capital for its own sake.

Of course, the effect of this was not felt in a day or even a century. The expansion had a cumulative effect. It takes time to cut down trees. The trees of Ireland were all cut down in the seventeenth century. But there were other trees elsewhere. Today we talk about the Amazon rain forest as the last real expanse, and it seems to be going fast. It takes time to pour toxins into rivers or into the atmosphere. A mere fifty years ago, smog was a newly-invented word to describe the very unusual conditions of Los Angeles. It was thought to describe life in a locale that showed a heartless disregard for the quality of life and high culture. Today, smog is everywhere; it infests Athens and Paris. And the capitalist world-economy is still expanding at a reckless rate. Even in this Kondratieff-B downturn, we hear of remarkable growth ratios of east and southeast Asia. What may we expect in the next Kondratieff-A upturn?

Furthermore, the democratization of the world, and there has been a democratization, has meant that this expansion remains incredibly popular in most parts of the world. Indeed, it is probably more popular than ever. More people are demanding their rights, and this includes quite centrally their rights to a cut in the pie. But a cut in the pie for a large percentage of the world's population necessarily means more production, not to mention the fact that the absolute size of the world population is still expanding as well. So it is not only capitalists but ordinary people who want this. This does not stop many of these same people from also wanting to slow down the degradation of the world environment. But that simply proves that we are involved in one more contradiction of this historical system. That is, many people want to enjoy both more trees and more material goods for themselves, and a lot of them simply segregate the two demands in their minds.

From the point of view of capitalists, as we know, the point of increasing production is to make profits. In a distinction that does not seem to me in the least outmoded, it involves production for exchange and not production for use. Profits on a single operation are the margin between the sales price and the total cost of production; that is, the cost of everything it takes to bring that product to the point of sale. Of course, the actual profits on the totality of a capitalist's operations are calculated by multiplying this margin by the amount of total sales. That is to say, the "market" constrains the sales price in that, at a certain point, the price becomes so high that the total sales profits is less than if the sales price were lower.

But what constrains total costs? The price of labor plays a very large role in this, and this of course includes the price of the labor that went into all of the inputs. The market price of labor is not merely, however, the result of the relationship of supply and demand of labor but also of the bargaining power of labor. This is a complicated subject with many factors entering into the strength of this bargaining power. What can be said is that, over the history of the capitalist world-economy, this bargaining power has been increasing as a secular trend, whatever the ups and downs of its cyclical rhythms. Today, this strength is at the verge of a singular ratchet upward as we move into the twenty-first century because of the deruralization of the world.

Deruralization is crucial to the price of labor. Reserve armies of labor are of different kinds in terms of their bargaining power. The weakest group has always been those persons resident in rural areas who come to urban areas for the first time to engage in wage employment. Generally speaking, for such persons the urban wage, even if extremely low by world or even local standards, represents an economic advantage over remaining in the rural area. It probably takes 20 to 30 years before such persons shift their economic frame of reference and become fully aware of their potential power in the urban workplace, such that they begin to engage in syndical action of some kind to seek higher wages. Persons long resident in urban areas, even if they are unemployed in the formal economy and living in terrible slum conditions, generally demand higher wage levels before accepting wage employment. This is because they have learned how to obtain from alternative sources in the urban center a minimum level of income higher than that which is being offered to newly arrived rural migrants.

Thus, even though there is still an enormous army of reserve labor throughout the world-system, the fact that the system is being rapidly deruralized means that the average price of labor worldwide is going up steadily. This means in turn that the average rate of profits must necessarily go down over time. This squeeze on the profits ratio makes all the more important the reduction of costs other than labor costs. But, of course, all inputs into production are suffering the same problem of rising labor costs. While technical innovations may continue to reduce the costs of some inputs and governments may continue to institute and defend monopolistic positions of enterprises permitting higher sales prices, it is nonetheless absolutely crucial for capitalists to continue to have some important part of their costs paid by someone else.

This someone else is of course either the state or, if not the state directly, then the "society." Let us investigate how this is arranged and how the bill is paid. The arrangement for states to pay costs can be done in one of two ways. The governments can accept the role formally, which means subsidies of some kind. However, subsidies are increasingly visible and increasingly unpopular. They are met with loud protests by competitor enterprises and by similar protests by taxpayers. Subsidies pose political problems. There is another, more important, way, which has been politically less difficult for governments because all it requires is non-action. Throughout the history of historical capitalism, governments have permitted enterprises not to internalize many of their costs by failing to require them to do so. They do this in part by underwriting infrastructure and in part, probably in larger part, by not insisting that a production operation include the cost of restoring the environment in such a way that it is "preserved."

There are two different kinds of operations in preserving the environment. The first is the cleaning up of the negative effects of a production exercise (for example, combating chemical toxins that are a by-product of production or removing nonbiodegradable waste). The second is investment in the renewal of the natural resources that have been used (for example, replanting trees). Once again, the ecology movements have put forward a long series of specific proposals that would address these issues. In general, these proposals meet with considerable resistance on the part of the enterprises that would be affected by such proposals, on the grounds that these measures are far too costly and would therefore lead to the curtailment of production.

The truth is that the enterprises are essentially right. These measures are indeed too costly, by and large, if we define the issue in terms of maintaining the present average worldwide rate of profit. They are too costly by far. Given the deruralization of the world and its already serious effect upon the accumulation of capital, the implementation of significant ecological measures, seriously carried out, could well serve as the coup de grâce to the viability of the capitalist world-economy. Therefore, whatever the public relations stance of individual enterprises on these questions, we can expect unremitting foot-dragging on the part of capitalists in general. We are in fact faced with three alternatives. One, governments can insist that all enterprises internalize all costs, and we would be faced with an immediate acute profits squeeze. Or, two, governments can pay the bill for ecological measures (clean-up and restoration plus prevention) and use taxes to pay for this. But if one increases taxes, one either increases the taxes on the enterprises, which would lead to the same profits squeeze, or one raises taxes on everyone else, which would probably lead to an acute tax revolt. Or, three, we can do virtually nothing, which will lead to the various ecological catastrophes of which the ecology movements warn. So far, the third alternative has been carrying the day. In any case, this is why I say that there is "no exit," meaning by that that there is no exit within the framework of the existing historical system.

Of course, if governments refuse the first alternative of requiring the internalization of costs, they can try to buy time. That is, in fact, what many have been doing. One of the main ways to buy time is to try to shift the problem from the politically stronger to the backs of the politically weaker, that is, from North to South. There are two ways in turn to do this. One is to dump the waste in the South. While this buys a little time for the North, it doesn't affect global cumulation and its effects. The other is to try to impose upon the South a postponement of "development" by asking them to accept severe constraints on industrial production or the use of ecologically sounder but more expensive forms of production. This immediately raises the question of who is paying the price of global restraints and whether in any case these partial restraints will work. If China were to agree, for example, to reduce the use of fossil fuels, what would this do to the prospects of China as an expanding part of the world market and therefore to the prospects for capital accumulation? We keep coming back to the same issue.

Frankly, it is probably fortunate that dumping on the South provides in fact no real long-term solution to the dilemmas. One might say that such dumping has been part of the procedure all along for the past 500 years. But the expansion of the world-economy has been so great, and the consequent level of degradation so severe, that we no longer have the space to adjust significantly the situation by exporting it to the periphery. We are thus forced back to fundamentals. It is a matter of political economy first of all and consequently a matter of moral and political choice.

The environmental dilemmas we face today are directly the result of the fact that we live in a capitalist world-economy. While all prior historical systems transformed the ecology, and some prior historical systems even destroyed the possibility of maintaining a viable balance in given areas that would have assured the survival of the locally existing historical system, only historical capitalism, by the fact that it has been the first such system to englobe the earth and by the fact that it has expanded production (and population) at a previously unimaginable rate, has threatened the possibility of a viable future existence for mankind. It has done this essentially because capitalists in this system succeeded in rendering ineffective the ability of all other forces to impose constraints on their activity in the name of values other than that of the endless accumulation of capital. It is precisely Prometheus unbound that has been the problem.

But Prometheus unbound is not inherent in human society. The unbounding, of which the defenders of the present system boast, was itself a difficult achievement, whose middle-term advantages are now being overwhelmed by its long-term disadvantages. The political economy of the current situation is that historical capitalism is in fact in crisis precisely because it cannot find reasonable solutions to its current dilemmas, of which the inability to contain ecological destruction is a major one, if not the only one.

I draw from this analysis several conclusions The first is that reformist legislation has built-in limits. If the measure of success is the degree to which such legislation is likely to diminish considerably the rate of global environmental degradation in say the next 10 to 20 years, I would predict that the answer is, very little. This is because the political opposition can be expected to be ferocious, given the impact of such legislation on capital accumulation. It doesn't follow, however, that it is therefore pointless to pursue such efforts. Quite the contrary, probably. Political pressure in favor of such legislation can add to the dilemmas of the capitalist system. It can crystallize the real political issues that are at stake, provided, however, that these issues are posed correctly.

The entrepreneurs have argued essentially that the issue is one of jobs versus romanticism or humans versus nature. To a large degree, many of those concerned with ecological issues have fallen into the trap by responding in two different ways, both of which are, in my view, incorrect. The first is to argue that "a stitch in time saves nine." That is to say, some persons have suggested that, within the framework of the present system, it is formally rational for governments to expend x-amounts now in order not to spend greater amounts later. This is a line of argument that does make sense within the framework of a given system. But I have just argued that, from the point of view of capitalist strata, such "stitches in time," if they are sufficient to stem the damage, are not at all rational in that they threaten in a fundamental way the possibility of continuing capital accumulation.

There is a second, quite different argument that is made, which I find equally politically impractical. It is the argument on the virtues of nature and the evils of science. This translates in practice into the defense of some obscure fauna of whom most people have never heard, and about which most people are indifferent, and thereby puts the onus of job destruction on flaky middle-class urban intellectuals. The issue becomes entirely displaced from the underlying ones, which are, and must remain, two. The first is that capitalists are not paying their bills. And the second is that the endless accumulation of capital is a substantively irrational objective and that there does exist a basic alternative which is to weigh various benefits (including those of production) against each other in terms of collective substantive rationality.

There has been an unfortunate tendency to make science and technology the enemy, whereas it is in fact capitalism that is the generic root of the problem. To be sure, capitalism has utilized the splendors of unending technological advance as one of its justifications. And it has endorsed a version of science—Newtonian, determinist science—as a cultural shroud, which permitted the political argument that humans could indeed "conquer" nature, should indeed do so, and that thereupon all negative effects of economic expansion would eventually be countered by inevitable scientific progress.

We know today that this vision of science and this version of science is of limited and not universal applicability. This version of science is today under fundamental challenge from within the community of natural scientists themselves, from the now very large group who pursue what they call "complexity studies." The sciences of complexity are very different from Newtonian science in various important ways: the rejection of the intrinsic possibility of predictability; the normality of systems moving far from equilibrium with their inevitable bifurcations; the centrality of the arrow of time. But what is perhaps most relevant for our present discussion is the emphasis on the self-constituting creativity of natural processes, and the nondistinguishability of humans and nature, with a consequence assertion that science is of course an integral part of culture. Gone is the concept of the rootless intellectual activity, aspiring to an underlying eternal truth. In its place we have the vision of a discoverable world of reality, but one whose discoveries of the future cannot be made now because the future is yet to be created. The future is not inscribed in the present, even if it is circumscribed by the past.

The political implication of such a view of science seems to me quite clear. The present is always a matter of choice, but as someone once said, although we make our own history, we do not make it as we choose. Still, we do make it. The present is a matter of choice, but the range of choice is considerably expanded in the period immediately preceding a bifurcation, when the system is furthest from equilibrium, because at that point small inputs have large outputs (as opposed to moments of near equilibrium when large inputs have small outputs).

Let us return therefore to the issue of ecology. I placed the issue within the framework of the political economy of the world-system. I explained that the source of ecological destruction was the necessity of entrepreneurs to externalize costs and the lack of incentive therefore to make ecologically sensitive decisions. I explained also, however, that this problem is more serious than ever because of the systemic crisis into which we have entered. For this systemic crisis has narrowed in various ways the possibilities of capital accumulation, leaving as the one major crutch readily available the externalization of costs. Hence, I have argued it is less likely today than ever before in the history of this system to obtain the serious assent of entrepreneurial strata to measures fighting ecological degradation.

All this can be translated into the language of complexity quite readily. We are in the period immediately preceding a bifurcation. The present historical system is in fact in terminal crisis. The issue before us is what will replace it. This is the central political debate of the next 25 to 50 years. The issue of ecological degradation, but not of course only this issue, is a central locus of this debate. I think what we all have to say is that the debate is about substantive rationality, and that we are struggling for a solution or for a system that is substantively rational.

The concept of substantive rationality presumes that in all social decisions there are conflicts between different values as well as between different groups, often speaking in the name of opposing values. It presumes that there is never any system that can realize fully all these sets of values simultaneously, even if we were to feel that each set of values is meritorious. To be substantively rational is to make choices that will provide an optimal mix. But what does optimal mean? In part, we could define it by using the old slogan of Jeremy Bentham, the greatest good for the greatest number. The problem is that this slogan, while it puts us on the right track (the outcome), has many loose strings.

Who, for example, are the greatest number? The ecological issue makes us very sensitive to this issue. For it is clear that, when we talk of ecological degradation, we cannot limit the issue to a single country. We cannot even limit it to the entire globe. There is also a generational issue. What may be the greatest good for the present generation may be very harmful to the interests of future generations. On the other hand, the present generation also has its rights. We are already in the midst of this debate concerning living persons: percentage of total social expenditures on children, working adults, and the aged. If we now add the unborn, it is not at all easy to arrive at a just allocation.

But this is precisely the kind of alternative social system we must aim at building, one that debates, weighs, and collectively decides on such fundamental issues. Production is important. We need to use trees as wood and as fuel, but we also need to use trees as shade and as esthetic beauty. And we need to continue to have trees available in the future for all these uses. The traditional argument of entrepreneurs is that such social decisions are best arrived at by the cumulation of individual decisions, on the grounds that there is no better mechanism by which to arrive at a collective judgment. However plausible such a line of reasoning may be, it does not justify a situation in which one person makes a decision that is profitable to him at the price of imposing costs on others, without any possibility for the others to intrude their views, preferences, or interests into the decision. But this is what the externalization of costs precisely does.

No exit? No exit within the framework of the existing historical system? But we are in the process of exit from this system. The real question before us is where we shall be going as a result. It is here and now that we must raise the banner of substantive rationality around which we must rally. We need to be aware that once we accept the importance of going down the road of substantive rationality, that this is a long and arduous road. It involves not only a new social system, but new structures of knowledge in which philosophy and sciences will no longer be divorced, and we shall return to the singular epistemology within which knowledge was pursued everywhere prior to the creation of the capitalist world-economy. If we start down this road, in terms of both the social system in which we live and the structures of knowledge we use to interpret it, we need to be very aware that we are at a beginning and not at all at an end. Beginnings are uncertain and adventurous and difficult, but they offer promise, which is the most we can ever expect.

2

The Horsemen and the Killing Fields: The Final Contradiction of Capitalism

Peter E. Grimes

INTRODUCTION

We live today in a time of unprecedented crisis on a global scale. This is a point of agreement shared by most scientists examining planetary trends. It is also a point many nonscientists sense intuitively. They show their fear in subtle but revealing ways: rising support for religious fundamentalism, ethnic separatism, millenarianism, and a generalized "hunkering down" into enclaves within which they feel "safe" against a dark and uncertain future. This popular sentiment is a murky reaction to real threats only occasionally referenced by the sanctioned media—erratic and violent weather; generation-long and global rises in the rates of cancer, inequality, and poverty; urban unemployment; shrinking government services. These are global and long-term trends from which no one is immune. Yet they are subject to local variations and reversals (such as the current drop in U.S. crime and unemployment rates), allowing for the transient illusion that one or another nation, region, class, or ethnicity may be safe. But these illusions only fuel motivation for the very separatism that can block effective global solutions.

The reality felt only dimly on a popular level is well known to the scientific community. Rates of economic growth fell worldwide between the 1970s and the early 1990s, while the biosphere continues to be shredded ever more efficiently by such growth as remains (Stevens 1997, June 17, C-8). Global warming, deforestation, collapsing fisheries, and ozone depletion are collectively combining to touch everyday life, while new versions of old diseases are reviving ancient plagues. At the same time, the contraction of high-wage jobs has cut into the tax base of governments across the core, encouraging them to cut social spending at the very time that it is needed the most.

The crisis is real, urgent, and global. Popular fear is warranted. But without correct information, that fear lends itself to manipulation by demagogues preaching isolation and separation for personal gain, thereby erecting barriers of fear to the very cooperation that is so necessary for common survival. Presented here is an effort to provide needed scientific information about our crisis in a clear, systematic, and accessible way, reaching for a unified analysis that links the elimination of nature with the economic deprivation of everyday life, while showing at the same time why the urge toward political separatism is both so tempting yet so collectively fatal.

Since the crises of the biosphere, economy, and political legitimacy are mutually interactive, the unraveling of their causal links is similar to teasing apart a knot in thread—all of the parts are connected, so the place to begin is almost arbitrary. Here we start at the ground and move up, both conceptually and ecologically.

EVOLUTION AND HABITAT

All life processes are driven by energy, and for the vast majority of organisms the source of that energy is solar, captured first by the plants and then sequentially consumed by herbivores and carnivores. Under typical conditions plants capture about 2 percent of the incoming sunlight, while herbivores and carnivores can at best access 10 percent of the energy stored in the bodies of plants and grazing animals, respectively (Bonner 1988; Colinvaux 1978). In these terms of energy flow, the various means by which human societies have been organized (e.g., hunter-gatherer bands, horticultural chiefdoms, agricultural empires, and the current structures of global capitalism) can be understood as ever more aggressive efforts to channel solar energy away from competing species and toward exclusively human consumption. The nested problems of our times can also be stated in these thermodynamic terms as arising from the collision of our expanding energy consumption with the limits set by primary plant production.[1]

The dependence of humans upon plant production has ultimately forced our species to relinquish its original freedom as roaming gatherers, scavengers, and occasional hunters in favor of securing a predictable future food supply as farmers, thereby cutting out the "middlemen" of herbivore insects and animals. The passage of the millennia since those initial settlements has allowed plenty of time for the development of agricultural techniques that maximized yield per acre, technologies specifically adapted to local conditions of soil and climate. One of our current problems lies with the misapplication of techniques developed for the temperate regions to the tropics. But to understand why this is, we must first digress into the question of soil types.

Climate, Plants, and Soils

The climactic stability of the period since the end of the last ice age has allowed for the stable reproduction of locally adapted plants, which give the appearance of being, in the words of one ecologist, "Nation-States of Trees" (Colinvaux 1978, chap. 5). Huge areas of continents worldwide are dominated by a narrow range of similar tree, bush, and grass varieties. Below the arctic tundra are hundreds of miles of conifers and other evergreens, merging as one moves south almost imperceptibly into deciduous hardwoods, themselves gradually giving way to either desert cacti or tropical broad-leafed softwoods, depending on the abundance of rain. Finally, of course, there is the broad band of tropical rain forest around the equator, the object of so much recent world attention.

Ecological investigation has demonstrated that these broad areas of plant similarity are the products of evolutionary selection responding to the climactic stability experienced within each of these different biomes: a stability of temperature, wind, and precipitation acting over the nine millennia since the end of the last glaciation (Colinvaux 1978, chap. 5). The key insight that explains these various plant forms is their efforts to maintain the conditions of temperature, sunlight, and nutrient flow that will optimize photosynthesis within the limits imposed by their local climates. Below the arctic treeline the dominant vegetation is evergreens. The needle shape of their "leaves" minimizes the heat loss of evaporative cooling while still allowing for a high density of chloroplasts (the site of photosynthesis), which gives them their dark green color. Further, this shape's thermal efficiency allows the needles to be productive of sugar energy even during very cold and/or cloudy periods.

South of the broad belt of conifers lies a contiguous belt of mixed conifers and deciduous trees. During the summers of this northern temperate zone, the broad-leafed deciduous trees are at a distinct advantage, gathering solar energy much faster than their conifer cousins. But during the winter, the conditions are exactly as in the arctic. Here the needle strategy excels over the broad-leafed, allowing the conifers to prosper while the broad-leafed plants have given up altogether, shedding their leaf factories and escaping into hibernation. When the line of arctic weather retreats northward in the spring, the deciduous broad-leafed strategy once again triumphs, and the broad leaves are generated anew. Further south, below the southernmost reach of the arctic winter, the high rate of evaporative cooling characteristic of the broad-leafed deciduous trees becomes an adaptive advantage because the summers are warm and humid. Under these conditions the conifer strategy no longer pays off, because needles become too hot in the summer and do not compensate adequately during the short and mild winters.

In the deserts, plant life takes on shapes that minimize the surface area exposed to the sun while maximizing the surface area exposed to wind—the exact opposite of the conifers of the North. Also, they have evolved means of carefully guarding their water against unnecessary loss. Yet they share with the tundra plants of the far North the quality of extremely slow growth, reflecting the severity of the struggle against their harsh climates, a struggle that allows only the most meager rate of biomass accumulation.

At last we come to the tropics, where weather is stable year-round except when punctuated by storms. As in the desert, the sheer abundance of sunlight requires some mechanism of cooling. However the copious supply of water allows for very broad-leafed plants to prosper, shedding excess heat by maximizing evaporative cooling. The absence of winter means both that the broad leaves can be retained permanently and that thick hard bark is unnecessary. All of the retained solar energy can thereby be released for growth, a condition that generates the profusion of biomass stereotypical of our images of tropical jungles.

Most of these plant designs had evolved long before the last ice age. But it has only been since the last ice age that they assumed their current geographic positions.[2] During the nine millennia since, they have changed the composition of the soils beneath them in fundamental ways, ways that continue to channel where and how we can grow food.

AGRICULTURE

Soil, Plants, and Social Structure

In the broad belt of conifers ringing the arctic, centuries of needle accumulation have led to acidic soils of limited agricultural value even in areas having a growing season. But further south, in the mixed boreal/deciduous forests and even more in the purely deciduous biomes, the fallen leaves have lower acidity. Further, the annual winters retard the decomposition process, allowing for the slow accretion of organic humus at the rate of about one inch per century (Colinvaux 1978, chap. 7). This organic residue is unique to the temperate zones and has gradually altered soil chemistry so as both to infuse it with nutrients and also to make it chemically more receptive to bonding with them. It is this same combination of soil ingredients unique to the temperate zone that allows for the irresponsible farmer to reuse the same plot of land almost indefinitely.

Further south still (skipping over the deserts) in the tropics, the majority of the soil is sterile. This appears bizarre when one considers the plush and abundant growth of a rain forest. The answer lies in the vigorous and competitive growth of life enabled by the constant moist warmth there. The surface life on top of the soil (fungi, bacteria, insects, and their predators) so quickly and efficiently devours fallen dead wood and animals that their nutrients never get the chance to get absorbed by the ground. Any molecular morsel remaining after this thorough treatment by decomposers is eagerly snatched by the network of near-surface roots supplying the forest trees. The evolutionary efficiency of this matrix of life in the rain forest (ultimately powered by the constant stream of intense solar energy) prevents the percolation of nutrients down into the soil, thereby disabling the chemical processes that sustain the gray-brown soils of the temperate zones. Instead, tropic and near-tropic soils are red, having long ago been washed clean by millennia of rain of the clay silicates that capture and reproduce the gray-brown humus of the temperate biomes.[3]

When our ancient ancestors eventually became compelled by population growth to abandon the hunter-gatherer life for the more predictable and controlled foraging and eventual planting of what would slowly unfold into horticultural production, they became tethered to particular locations, their ranges constricted by the requirements of crop maintenance. This "neolithic revolution" can be reconstructed by the artifacts recovered by archeologists and allows us to locate and date the earliest settlements. The earliest evidence dates from around 9–7,000 years B.P. (7–5,000 B.C. or almost immediately after the end of the "ice age"), and the majority of locations lie between 20° and 40° north of the equator (Sanderson 1995, 112–120).

Abundant evidence suggests that hunter-gatherer bands were well aware of the technology of horticulture long before they chose to use it, presumably because they knew also that it would require much more time and energy than they were willing to invest. But, eventually, growing population density in the temperate regions required the shift to plant cultivation in order to reduce between-group competition and warfare.[4] The same processes also operated in both the tropics and the polar regions, but the much poorer land productivity there precluded the solution adopted in the temperate realms of increasing land productivity: instead competing groups were compelled to separate from each other across much greater spaces in order to survive or else be condemned to constant warfare (Chase-Dunn and Hall 1997).

For the peoples of the north polar regions living atop or just south of the vast ice sheets capping the arctic, exploitation of plant energy was never a serious option. Instead, they were compelled by the constraints of their climate to live as carnivores, searching out prey (marine and terrestrial) whose migrations north served as imports of solar energy from the south. Hence the development of complex societies based upon dense human settlements was vitiated from the start.

Similar constraints operated in the tropics. This may at first seem strange. After all, it is the tropics that have always had the strongest and most consistent input of solar energy leading to the greatest biodiversity and may well have been the initial environment of the first humans. Yet most of the areas of initial neolithic settlement are well north of the tropics.[5] Once again, the answer lies in the soil.

Among indigenous peoples still living in the tropics, rotating slash-and-burn (swidden) horticulture continues to be the technology of choice. This is because the ashes left from burning temporarily boost the fertility of the soil, allowing for the growth of edible plants until the soil is again washed clean by the rains. Once harvested, the cultivated area is deliberately allowed to revert to forest. Over eons this strategy has worked, because the small size of the plots of land involved is well within the scale that can be eventually repopulated by the limited colonization strategies available to vegetation in the tropics. However, the ecological constraints placed on the reproduction and spread of tropical plants places a strict upper limit on the density of human population that can be supported by swidden technology. This ecological barrier—as rigorous as that operating in the polar regions—here again prevented the emergence of the densities of human settlement required for complex social systems, forever stalling the development of societies more complex than chiefdoms in the tropics.

These preconditions of long-term soil fertility are found only in the soils beneath the temperate forests. Hence the neolithic breakthrough to collectively managed horticulture on a large scale was both compelled by the population density accumulating in favorable climates, while yet also being enabled by the combination of solar energy and soil fertility peculiar to those climates. This technical and organizational revolution was also the first historical demonstration of an ability perhaps unique to humanity—the capacity to collectively plan and execute a strategy to deliberately and systematically divert solar energy away from competing life-forms toward human use.[6]

The soils laid down and evolving under the deciduous broad-leafed regions north of the tropics were of the gray sort most fertile for cultivation, which had been gradually stocking up precious topsoil at the rate of one inch per century. These were the soils that became the ecological foundation for the complex "tributary"[7] empires of our past—the Mesopotamian civilizations; the Harappan, Mauryan, and Gupta civilizations of India; the Han dynasty of China; the Greco-Roman empires. But although more tolerant of human agricultural exploitation than other soil types, they can still be exhausted by sufficient misuse. The current arid sterility of the Middle East and much of the peninsulas of the Mediterranean bears quiet testimony to the depredations of past abuse: massive deforestation (allowing erosion of topsoil) along with grossly excessive irrigation (salting the soil to toxic levels). Recent research suggests that the collapse of more than one empire may have been rooted in soil depletion (Runnels 1995; Chew 1996). But despite these excesses, the greater fertility of the bulk of the temperate regions eventually came to support the highest population densities on the planet throughout the nine millennia since the neolithic revolution.

CAPITALIST TECHNOLOGY AND GLOBAL AGRICULTURE

Machinery and Fossil Fuels

Capitalist production relations in the agricultural sector did not of itself immediately change the technology of food production refined during the centuries of the tributary mode.[8] Europe had, by the eighteenth century, long been familiar with the medieval use of the three-field rotation system that restored soil fertility by the growth of nitrogen-fixing plants (often legumes or clover) as one of the three crops; Southeast Asia had likewise settled the technical aspects of optimum production for rice paddies; while the Andean peoples of the Incan civilization had devised ingenious methods of prolonging the growing season for potatoes by planting them in raised rows adjacent to troughs of water. Each of these approaches to maximizing land productivity had evolved over centuries of experimentation in their respective regional biomes, and the development (or imposition via conquest) of capitalism did not change these approaches.

However, the adaptation of steam power to the development of the first tractors near the turn of the twentieth century was one of the very first applications of the current era of technological revolution to a rural environment. While designed originally as a replacement for the horse, the early tractors were too expensive for purchase by the typical family farmer. So instead entrepreneurs would buy them cooperatively while also hiring men to operate and maintain them and send these teams of men and machines from farm to farm for hire. The beginning of this new technology was in the United States, both because that was where the first manufacturers of this equipment were located and because farmers in the U.S. were relatively wealthy as compared to their counterparts elsewhere on the planet.[9] As the market for farm machinery expanded, economies of scale, Taylorism,[10] and assembly-line organization allowed the price of farm machinery to drop. Immediately affected was the use of animals as motive power: between 1865 and 1915 the number of horses and mules for every hectare of cropland fluctuated around 0.3, with a high of 0.35 in the first year and a low of 0.25 in 1880. (These numbers and the ones following are all derived from figures provided by Mitchell [1993] and the U.S. Bureau of the Census.) But after 1915 (five years after the appearance of the tractor in the records) the number began to drop consistently, passing below 0.2 in 1930, 0.1 in 1950, and disappearing entirely from the books in 1965 (see Figure 2.1). To this trend out for animals we see the inverse for tractors. Starting with their appearance in the census records in 1910, the number of tractors per hectare planted in food crops doubled every five years until 1935, after which it shot up dramatically from 0.01 in 1935 to 0.08 in 1970. These numbers provide a very clear demonstration of the displacement of solar (organic) energy by fossil fuel (machine) power.[11] In addition, when we remember to include fertilizers as themselves drawn from fossil fuels, their application to cropland has closely followed the use of tractors, again shooting up fastest after 1935 (Figure 2.1). The development of machines also began to displace human labor as well. In 1830, agricultural workers comprised 70 percent of the workforce and in 1870 50 percent, but as early as 1913 the percentage was cut almost in half (27.5 percent), while by 1950 it had shrunk to 13 percent, and in 1991 was only 1.6 percent (Maddison 1995, table 2-5, 39; World Bank 1995b, table a-2, 148). Meanwhile, the share of U.S. agricultural production as a percentage of its GDP stayed largely the same between 1950 and 1990 (World Bank 1982, 1995b).

The substitution of mechanical energy powered by fossil fuels for the traditional organic labor of animals and people in all areas of production has been the most important trend of our era. But in agriculture as in all other spheres, the transition has entailed important costs. Here this cost is most obvious in the loss of topsoil that came with the new machinery.

Figure 2.1
Power Inputs per 1000 Crop Hectares in USA, 1865–1970

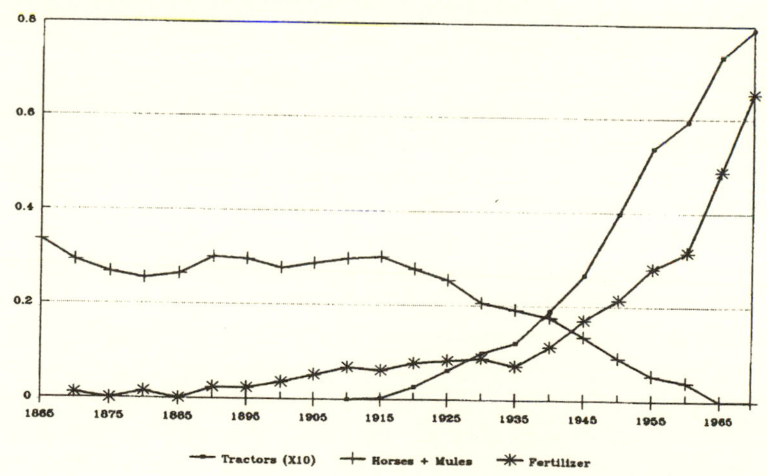

To understand why agricultural machinery has accelerated topsoil depletion we must first spell out some of the side effects of machinery itself. The most efficient application of large machines is to single-crop fields that are large. Otherwise the farmer sacrifices the economies of scale enabled by machinery to time spent changing the location and types of equipment. Animal-drawn ploughs can steer around large rocks and trees, so low-capital family farms relying on solar energy could function profitably along the east coast up into the Appalachian Mountains during the last century. But heavy equipment works best when unobstructed by rocks, trees, or bushes, which has led both to the westward migration of farming within the U.S. out of the mountains and into the plains states and also to deforestation and major loss of species diversity in the plains states. Meanwhile, ironically, this same movement west has allowed for a reforestation of the U.S. east coast and eastern Midwest as family farms have gone bankrupt (particularly Black-owned farms during the Depression), along with the return of birds, deer, coyotes, and bears to these regions (Stevens 1997, June 10, C-1; Rudel and Chun 1996). Loss of trees and non-crop plants in active farming areas eliminates windbreaks and greatly accelerates erosion and topsoil runoff, while the machinery itself cuts deeper into the ground than animal-drawn ploughs used to. (That's one reason why the Mississippi is also called the "Big Muddy," as flooding and erosion flush out the best soil and the chemicals applied to it from the midwest plains.) Another spin-off problem is that the water runoff takes not just topsoil but also the other chemicals applied to fields. Just as DDT killed birds, so a growth regulator found in vitamin A and used as a pesticide is currently thought to be a leading suspect behind a massive problem of gross frog mutation recently identified in Wisconsin (*All Things Considered,* National Public Radio, May 9, 1997).[12]

The mechanization of monocrop agriculture, when combined with the application of some kinds of petrochemicals to enhance soil fertility, other kinds to dampen competition from weeds, and still others to decrease predation by insects or disease, succeeded spectacularly in its original goals. Yield per acre and per worker both went up steadily throughout the twentieth century, and especially after World War II. These technical changes also achieved the goal of lowering consumer prices. But there have been other, less well-advertised, effects as well. The use of these techniques has become increasingly expensive, raising the barriers to entry even as they have reduced the price—and thus the profit per acre—of farming. An important result has been the consolidation of an increasingly corporate ownership along with the gradual washing out (accelerated during recessions) of family farms. From a low point of an average of 140 acres per farm in 1880, the mean size stayed roughly the same at below 150 until 1935. Then, mimicking the other indicators, it starts a sharp slope upwards of 25 to 50 acres every five years, reaching 461 in 1990 without showing any sign of stopping (Figure 2.2). Hence yield per acre rose consistently starting in the depression years, but did so at the cost of a total commitment to fossil fuels, artificial fertilizers, and the corporatization, automation, and centralization of ownership and production. The absolute number of people on farms peaked also in 1935 at around 31 million, after which it declined in an inverse echo of the other trends to a mere 10 million in 1970 (Figure 2.2). The same is true at an even higher level among the main beef suppliers: McMichael (1996, 102) reports that three corporations headquartered in the United States (Cargill, ConAgra, and Tyson's Foods), in cooperation with the Japanese firms C. Itoh and Nippon Meat Packers, supply a controlling share of global feedlot supplies and meat products.

Figure 2.2
Farm Population and Average Farm Size in USA, 1850–1990

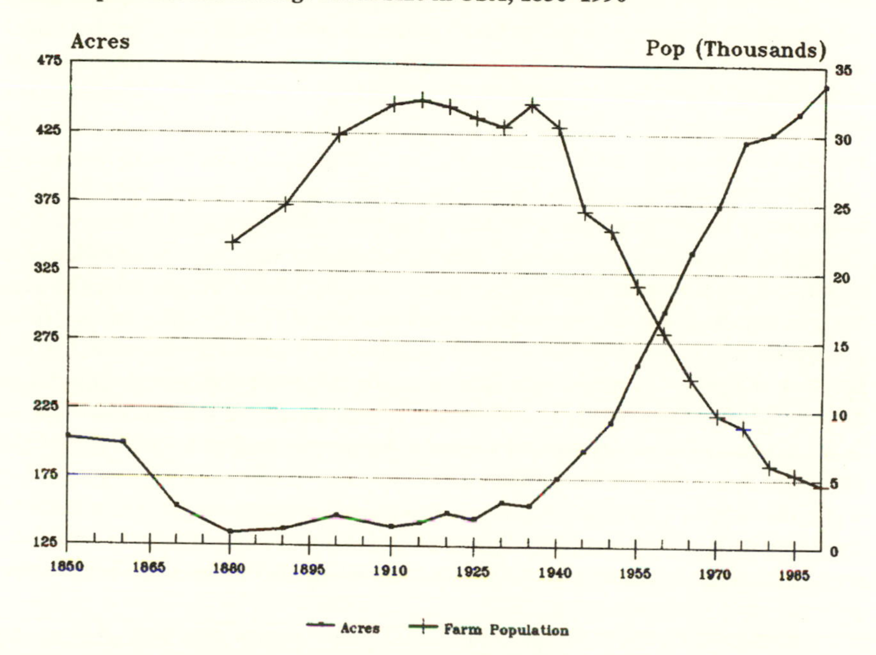

A Global Model: The "Green Revolution"

The detailed focus on the agricultural pattern inside the United States is not intended to be parochial. To the contrary, I have lingered on this example precisely because it was to become an intentional global model encouraged by U.S. policymakers for emulation throughout the world (McMichael 1996). As early as the Marshall Plan, the U.S. approach to agricultural production was being promoted by Washington to selected European farmers (*All Things Considered,* National Public Radio, June 5, 1997). During the following decades, the U.S. method of high fossil-fuel inputs was also extended to the periphery as the most efficient solution to the demands of the global agricultural market (McMichael 1996). The centerpiece of this campaign was the "Green Revolution" (a presumed prophylactic for the spread of the dreaded "Red" revolutions then breaking out across the periphery). It was an effort at a technical "fix" (increased food production) to a social problem (class polarization and generalized malnutrition).

The Green Revolution sought to increase the yields of rice and other staple crops for use in the periphery, along with building their resistance to predation by insects, fungi, and bacteria. In many ways it succeeded (and continues to evolve). However, optimal use of these altered varieties led to several unanticipated consequences.

For many of the crops "improved" by breeding (particularly during the initial periods of the 1960s and early 1970s), more water and fertilizer were needed than that traditionally applied to optimize performance. In the case of water, irrigation pumps were needed often enough that the introduction of these new crops had the effect of raising the economic barriers to entry, along with the minimum land area necessary to profit, thereby expelling poorer peasants to the cities (just as U.S. "family" farms have themselves been expelled) (Perlman 1977; Burbach and Flynn 1980). Alternatively, peasants pushed out of traditional land have colonized the poorer soils along the mountainsides by cutting down rain forests. Increased irrigation also accelerated the rate of topsoil loss and salinization, resulting in a long-term postwar decline of productive land area (Pimentel and Harvey 1995). The requisite fertilizers themselves had to be bought from the U.S. or other core countries (increasing the need for foreign exchange [U.S. dollars] and thereby also the pressure to grow cash crops for export instead of food), compounding rural class polarization. For peasants without enough money to buy new equipment, the solution nearest to hand has been to have more children to increase family labor power. This has been another unanticipated consequence of the "Green Revolution": delaying the completion of the demographic transition by prolonging the period of high fertility in the countryside. By tying peasant income directly to the world market price for their cash crops, lower and uncertain sales from the use of the new crop varieties sustained high rates of fertility (to create new labor input), despite the drop in mortality brought about by new public health measures (Folbre 1977; Grimes 1982; Mamdani 1972).

Further, the replacement of traditional plant species and their genetic diversity by an imported group dependent on further imports is a questionable long-term strategy. As the land becomes less fertile with degradation it eventually is abandoned. Its productive life thereafter can only be extended by growing cocaine or other very high-priced crops, because traditional crops do not even repay the cost of the fertilizer.

The overall result of machinery, irrigation, and petrochemicals applied on a global scale has been the corresponding amplification of the same topsoil loss and land degradation that characterizes the United States. The International Commission of Scientific Unions issued a report (1997) which concludes that, worldwide, 75 billion metric tons of topsoil are washed away annually by the combination of machinery and irrigation, which corresponds to .5 inch per year (recall that .5 inch represents the work of 50 years of deposition, which means that, should this rate continue, only 20 more years will eliminate all of the topsoil accumulated in the temperate zones since the last ice age, 10,000 BP). In his encyclopedic summary of global environmental issues, Caldwell (1996, 257–58) affirms that:

Possibly the most serious natural resource problem now affecting all nations is land degradation . . . many activities in traditional as well as in modern industrial society contribute to the deterioration of soil quality and the loss of agricultural land; losses include soil erosion, loss of fertility, laterization, salinization and water-logging, desiccation, conversion to urban uses, and contamination by toxic wastes . . . the effects of soil mismanagement are characteristically slow, incremental, and cumulative, so that internationally significant injury may not be evident until irreversible damage has been done.

Bundled into this transition is also a dangerous dependence on fossil fuels for most inputs to compensate. But, unlike the temperate soils that dominate the United States, a large part of the soils in the periphery are tropical, hence (once large sections are cleared) they quickly become completely dependent upon external inputs (Colinvaux 1978, chap. 7). Hence any prolonged disruption (or significant price increase) of imported inputs would necessarily result in a dramatic contraction of output and the transformation of the formerly tropical soil into a wasteland (it cannot revert to tropical forest because of the soil sterility of tropical soils discussed above [Colinvaux 1978]).

Crisis in Fresh Water

A closely related problem emerging in recent years has been a growing shortage of fresh water (Caldwell 1996, 258; la Riviere 1990, 37–48). The aspect of the global water cycle of concern here is the rate of flow. Fresh water on land is renewed by ocean evaporation (and desalinization) followed by rain over land, after which it eventually returns to the sea. Over geological time, fresh water has accumulated in glacial snow packs and underground aquifers. (A huge example of the latter is the Oglala aquifer—named after the Sioux tribe—which stretches from the Dakotas south as far as Kansas and Colorado.) The demand for fresh water for both irrigation (currently 70 percent of global demand [World Bank 1997]) and urbanization has come to exceed the flow provided by rain, most severely in drought-prone areas. To compensate, deeper wells have tapped aquifers. The Oglala aquifer has been tapped to supply Las Vegas, Los Angeles, and farms in southern California to augment the flow of the Colorado River (itself so drained that during some summers it no longer makes it to the ocean). In the Middle East, water access is an important obstacle to peace talks, and rationing is in effect along the Gaza strip. In the former Soviet Union, the Aral Sea has contracted 50 percent, and the remainder has dangerous levels of salinity and petrotoxins (BBC, *Outlook*, May 14, 1997). Over the short term, the retreat of glacial snow packs adds to river flow in more temperate climates, but that is at best a mixed blessing. We are collectively consuming our water "capital," which will ultimately require restoration of the balance via a massive contraction of use. This can only mean sharp contractions of agricultural output and urban size or use (la Riviere 1990, 37–48).

In the first few millennia of human agriculture, it was not understood that continuous irrigation eventually deposited enough salt on the soil surface that fertility disappeared. In an analogous fashion, only now is it becoming also clear that continuous irrigation from wells liberates arsenic from its bonds to the soil, creating a gradual buildup of arsenic in the well water. Arsenic is a cumulative toxin for which there is no known cure. Recently the BBC reported that the British Geological Survey has discovered that the problem has become so widespread in Bangladesh and parts of India that an estimated 30–60 million people are being poisoned by their well water, a problem sufficiently grave that the World Bank has dispatched a team to investigate (BBC, *The World Today,* May 8, May 19, 1997).

Put simply, current technologies used in global food production have achieved their historic highs of yield per acre only by supplementing natural energy inputs with ever-larger amounts of fossil fuel. Insofar as there are limits to the supply of fossil fuels, the enormous subsidy they provide must eventually grow smaller and finally stop altogether. By itself, that will lower yield per acre and thereby raise prices. Added to all this are the current uncertainties of global warming (explored at greater length below). If the warming reaches the levels now officially projected by the Intergovernmental Panel on Climate Change (IPCC 1990; IPCC 1992; Karl, Nichols, and Gregory 1997), then the reversion of arid farmland to desert (as is now already underway in the state of Nebraska) will tend to accelerate, illustrating the removal of marginal land from production. This will become another pressure acting to increase food prices.

In sum, the post-war explosion in global food production has been predicated on the extension to the tropics of a technology developed for application to the temperate zones. In both regions the technology requires the massive subsidy of a finite resource—fossil fuels—to boost production. While this subsidy is unsustainable over the long term in either region, it is particularly unsuitable to the tropics, where the baseline soil fertility is so very low that withdrawal of fossil fuels would quickly lead to complete agricultural collapse. To the degree that global food output is relying on a transient and artificial fertility of tropical soils, then to that same degree it is hostage to the availability of cheap fossil fuels.

INDUSTRIAL AUTOMATION AND THE BIOSPHERE

The "industrialization" of agriculture was, as the term itself implies, the transfer of technologies originally developed for application to urban manufacturing facilities to the growing of plants. Abundant research has established that the motivation for the development of machinery to production was to better control labor and reduce the number of employees where possible (e.g., Braverman 1974; Marglin 1974; Stone 1974). As was true of agriculture later, the result was a substitution of fossil fuels for solar (human) energy. Again as with agriculture, the economic results were spectacular. Maddison (1995, 36) has calculated that, in the United States, the value in constant 1990 dollars of just machinery and equipment alone per worker traces a path like that found in agriculture: while in 1820 the value was 281, in 1870 it had quadrupled to 1,367; in 1992 it was 39,636 (Figure 2.3). The clever application of machinery to an ever-expanding range of human activities has been geometric, and today it has been given an added boost by the microprocessor. Additional data Maddison provides demonstrates that this sequence of jumps in worker productivity found in the U.S. is typical throughout the world-economy, although the values are predictably smaller in the periphery (Figure 2.4) (Maddison 1995, 249). Once again, machinery powered by fossil fuels is both faster and more reliable than human workers powered by the foods grown in the sun. Yet once again there are unpleasant consequences not popularly understood.

Figure 2.3
U.S. Labor Productivity: Equipment Value Per Worker (1820–1990) in 1990 Dollars

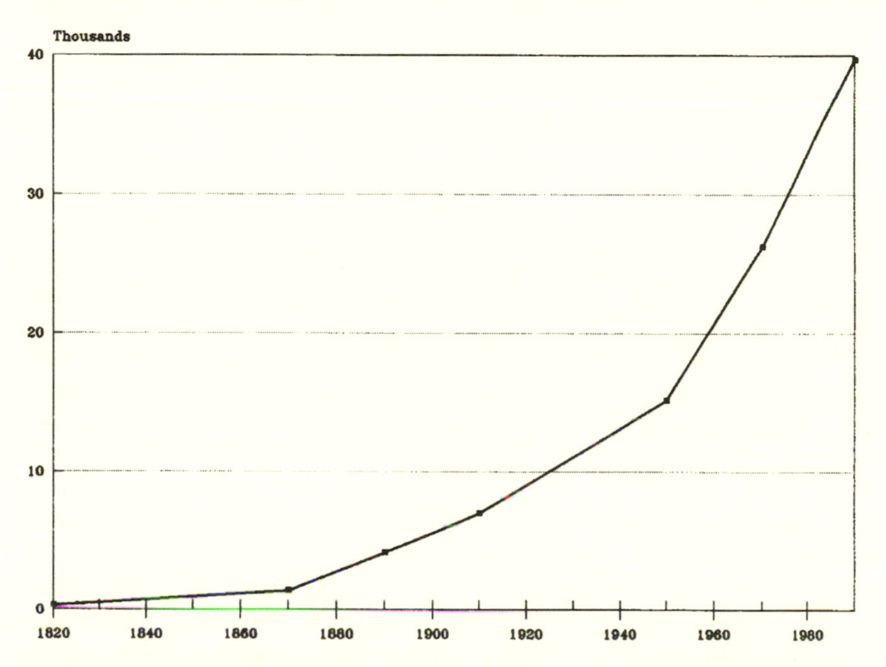

From 1930 to 1990, the global production of fossil fuels rose from 1.38 billion barrels of oil equivalent to 7.34 billion barrels; of which the United States produced 46 percent in the first year and 20 percent in the second (Figure 2.5; data from Etemad and Luciani 1991; the United Nations Energy Statistics Yearbooks). Inside the U.S. in 1987, 36 percent of its consumption was for industrial production and another 37 percent for transportation (mostly also connected with the requirements of production) (Gibbons, Blair, and Gwin 1990, 94). The upward burst of fossil fuel production after 1935 both globally and inside the U.S. parallels the history of the application of fossil fuels to agriculture and industry documented here for the U.S. alone.

Among the unwelcome by-products added to the biosphere by recent industrial production technologies have been excessive heat, resource exhaustion, acid rain, environmental estrogens, chlorofluorocarbons (CFCs) which cause ozone depletion, and greenhouse warming. We will take these in turn:

Figure 2.4
Labor Productivity: GDP Per Hour, Selected Countries (1870–1990) in 1990 Dollars

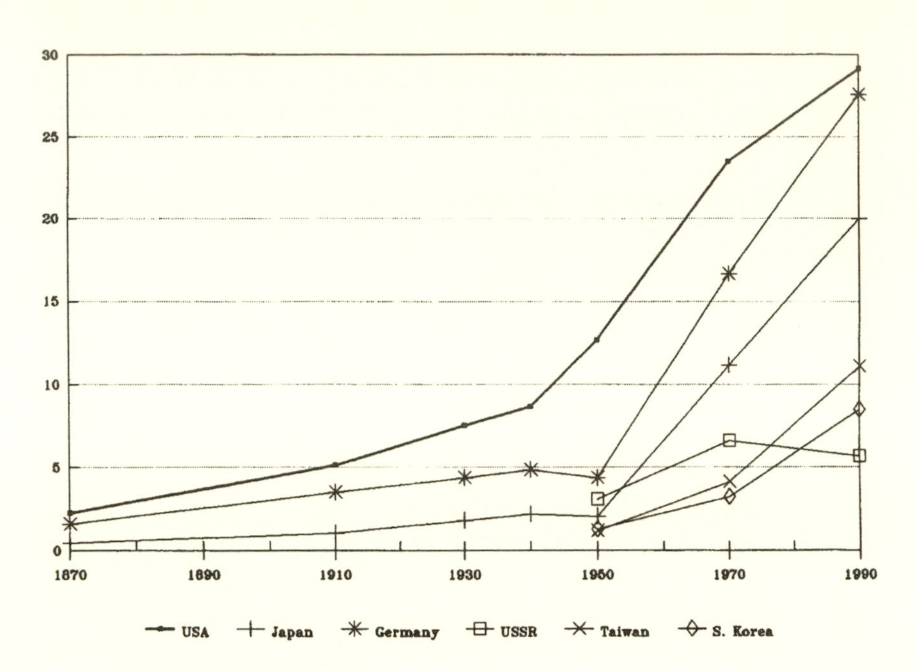

Entropy and Heat

The simplest and most elemental spin-off of any material transformation is heat. The second law of thermodynamics requires that all material transformations involve energy, and the greater either the speed of those transformations or their degree, the greater the energy involved. Energy, in turn, ultimately degrades into its lowest form—heat. It is this that, along with the heat absorption and low reflectivity of many roads and building materials, accounts for why most large cities average 10° (F) warmer than the surrounding countryside. This simple fact is the ultimate brake on industrial production. If there were no economic barriers to the conversion of earth's available materials into machines, tools, and the energy to run them sufficient to supply today's global population at the living standard that was typical of the core in the early 1970s, the energy required for this massive transformation of matter would by itself generate enough heat to turn the atmosphere into an oven that would cook us all to death (Commoner 1977; Georgescu-Rogen 1971).

A second, less often articulated, implication of the thermodynamics of production is the dispersion of mineral resources. Iron ore, for example, is initially concentrated in a mining site. After purification and further concentration in a steel mill, it becomes a component in commodities that are distributed globally. When those goods are discarded by their end users, they eventually rust, leaving behind a small area of iron oxide that returns to the soil. The net effect of this process is

Figure 2.5
Fossil Fuel Production: U.S. and World, 1800–1995

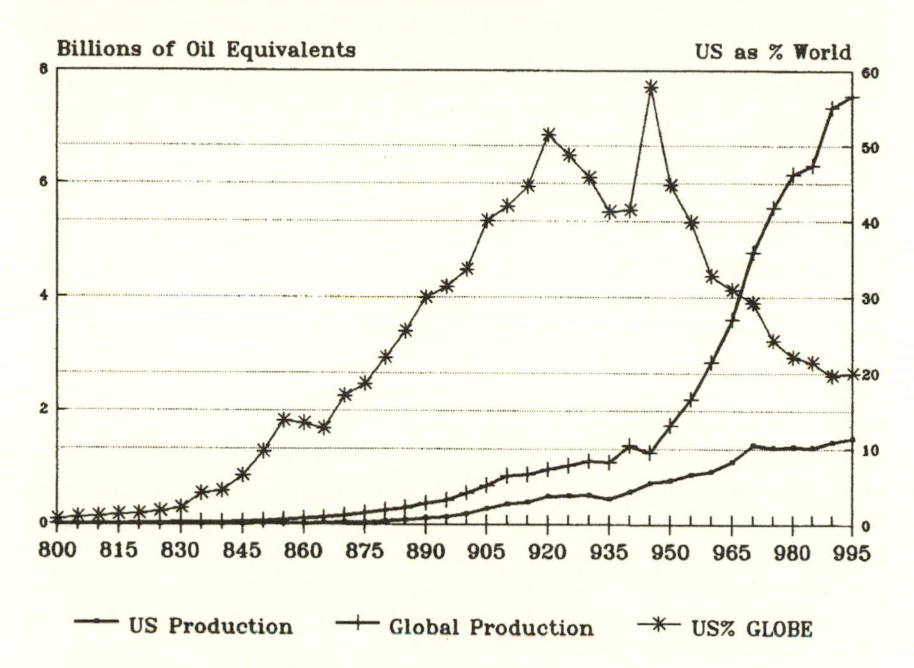

that resources initially found in abundant and concentrated pockets become dispersed in a way that makes them impossible to easily reclaim in the future. The overall volume remains the same, but its geographical redistribution fundamentally changes its accessibility to future generations.

Acid Rain

Acid rain results from the burning of coal with a high sulfur content. This coal, being more abundant and cheaper than the alternatives, is the fuel of choice for power plants throughout the world (particularly China). When burned, the exhaust contains sulfur dioxide (SO_2) which quickly combines with water in the atmosphere to create sulfuric acid (H_2SO_4). The rain downwind raises the acidity of soils and streams, killing trees, bacteria, and insects. Damage from acid rain has been well documented in the northeast of the U.S. and Canada, the Smokey Mountains of the southeastern U.S., and the Black Forest of Germany.

Environmental Estrogens

The production and use of plastics and certain electrical components has entailed the use of certain chemicals (e.g., PCBs) that are "loose cannons" in the ecosystem. They are collectively called "environmental estrogens" because they mimic natural estrogens and thereby may potentially interfere with the sexual evolution of a large number of different species (e.g., mature male alligators in Florida with immature testes unable to generate viable sperm, accelerated rates of sexual maturation of human female children [Painter 1997, 1]). Some have speculated that these chemicals may even be playing a role in the observed disappearance of frogs and amphibians (Blaustein and Wake 1995).

Ozone Depletion

More commonly known and now actually regulated are the effects of CFCs on the ozone layer. Used until recently as coolants in refrigerators and air conditioners, solvents for cleaning electrical components, and propellants for spray cans, CFCs were quite abundant in the 1960s and 1970s. After use, they eventually float up into the stratosphere, several miles up and well above the cloud layer. At that high level, ultraviolet (UV) light (very high-energy photons just beyond the blue-purple part of the visible range) from the sun floods in directly. Photons with these energies can easily strip electrons from atoms and break apart molecules, so they are extremely dangerous to life. Any human exposed naked to such light would quickly burn to death from the radiation. Under normal conditions, oxygen (O_2) hit by UV light splits in two (O_1) and then recombines into O_3—ozone. (The same thing happens when lightening strikes, giving air its distinctive "fresh" smell after a summer thunderstorm—a fact taken advantage of by car manufacturers who spray their products with ozone to give them that "new car" smell). Ozone has the capacity to absorb UV light and re-emit the absorbed energy in a more benign lower frequency form. CFCs interfere with this process by preferentially attracting the oxygen atoms split by UV, thus preventing their recombination into ozone. Hence the ozone hole first noticed above the Antarctic and now also perceptible as patches of depletion in the Northern Hemisphere.

The removal of the protection of ozone has allowed a dramatic increase of UV light proceeding unimpeded to the ground, especially in the portions of the Southern Hemisphere nearest the pole: South Australia, Chile, Argentina, and New Zealand. The effects have been observed in a myriad of forms, as an upsurge in skin cancer, cataracts, and blindness of sheep in New Zealand and kangaroos in South Australia. As with estrogens, some have speculated that excessive UV may be playing an additional role in the disappearance of the amphibians by sterilizing their eggs. Cases of melanoma have been proliferating in the U.S. and Canada as well, and followers of newspaper weather sections have doubtless noticed that most now feature a UV index as a routine part of their forecast. During the summer of 1995, the staff of the Baltimore Aquarium rescued a blind sea turtle from the coast of Delaware and were puzzled by the fact that he had cataracts while yet a juvenile of only 15 years. A less well-known outcome of increased UV radiation at ground level is a probable increase in mutations, especially among viruses, whose only protection from the air is a thin jacket of protein. The long-term results of such mutations are unpredictable.

The UN-sponsored international agreement on regulation and eventual elimination of CFCs (The Montreal Protocols reached in 1987) is one of the few true success stories of cooperation on a global level to address a global problem. Even so, it may take up to 50 years for the ozone layer to fully recover.

GLOBAL WARMING

Planetary Thermodynamics

Global warming from the introduction of greenhouse gases is another result of current technology found within both production and consumption. CFCs and methane (the latter largely from cattle) are each potent greenhouse gases, CFCs all the more so because they are very stable and very effective at blocking infrared (heat) radiation heading back out into space. However, the most abundant greenhouse gas is carbon dioxide (CO_2), a direct product of fossil fuels. All automobiles, planes, and ships burning hydrocarbons emit CO_2 along with all coal or oil-fired electric generators. Hence almost all of the machinery used in production contributes to CO_2 emissions.

An extreme outcome of heat entrapment from CO_2 and other greenhouse gases can be found on our nearest planetary neighbor Venus, whose proximity to the sun and heavily carbonized atmosphere sustains a surface temperature above the melting point of lead (over 700° F). Here on Earth, if it were not for naturally occurring levels of atmospheric CO_2, the surface temperature would be below freezing (0° F), liquid water would not exist, and life would be impossible. But it is a delicate balance. Too much warming and life (as we know it now) would die.

That temperature varies with the abundance of CO_2 is also clear from the fossil record and Antarctic ice cores. Those same ice cores reveal that levels of atmospheric CO_2 have risen 27 percent during the period 1800–1990, from 280ppm to 355ppm (Figure 2.6). Within the past two years (1994–1996), the Intergovernmental Panel on Climate Change (representing scientists from almost every country) has confirmed that global warming is already underway. What remains unknown is how far the process will go. Adding to the difficulties of prediction is the question of the missing carbon: calculated emissions of CO_2 have outpaced observed levels since reliable data on the former have been available (1950). Where is the missing carbon going? Several answers have been proposed, including root systems and ocean absorption. Whatever the specific cause of these natural carbon "sinks," it is likely that they will eventually fill up, in which case levels would catch up quickly with emissions, accelerating and intensifying the entire process.

Rising Sea Levels

It has already been observed that glacial ice packs are retreating while plants and butterflies have been documented moving higher up mountains and further north (Peters and Lovejoy 1990). Melting ice suggests rising sea levels, which satellite data now confirm. The retreat of the last ice sheets, which at their maximum 25,000 years ago stretched out across most of Europe, Russia, and North America in the North and equivalently in the South, raised sea levels over 300 feet.

Figure 2.6
Observed Levels of Atmospheric CO_2, 1790–1990

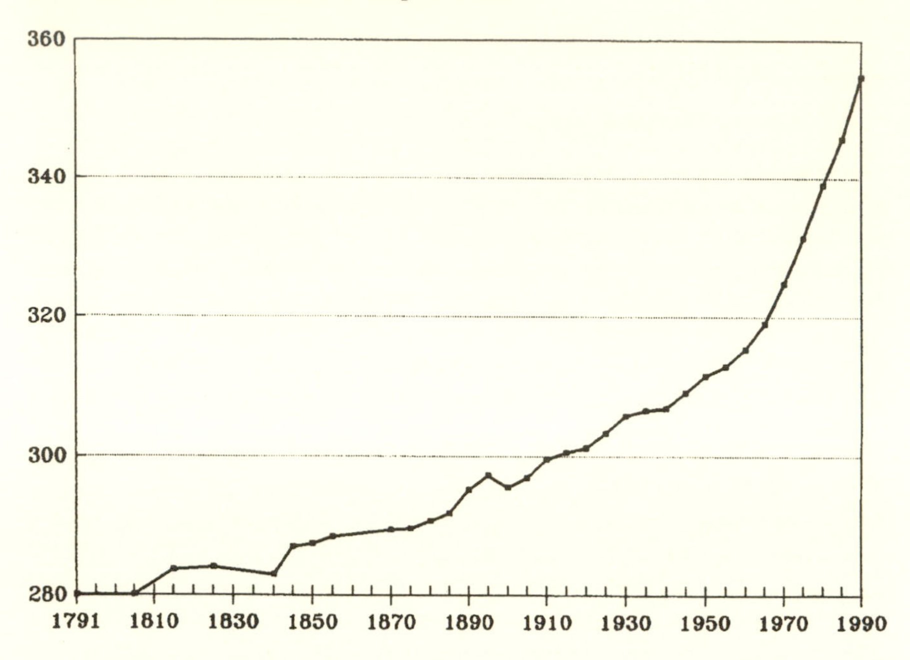

The remaining ice, if fully melted, would add another 250 feet. While we are yet far from that point, a rise of only one or two feet would permanently flood the current arable land around the Nile and has been estimated to be able to cut agricultural production globally by as much as 20 percent. Further, rising sea levels pose the potential for flooding important ports and coastal cities, as well as Pacific island states.

Disease

The zones supporting different forms of plant life also support connected bacteria, fungi, and viruses. As warming allow these to move north, diseases typically associated with the tropics (e.g., malaria, dengue fever) will follow into the areas now occupied by the countries in the temperate core.[13] Exacerbating this concern is the evolutionary mechanism of virulence. Diseases require hosts, and the tendency is for evolution to select for those diseases that can reside inside their hosts without killing them for long periods of time, long enough at least for them to survive until contact with a new host is possible. This evolutionary mandate is strongest where the host population is lowest. Conversely, an abundant supply of densely populated hosts removes that mandate, allowing for the transient flourishing of exceptionally virulent diseases that can kill quickly while still assured of transmission to new hosts. Just as the swine flu pandemic of 1918 is now suspected of having evolved under the unusually high concentrations of people in the trenches of World War I, so the extraordinary population densities in cities throughout the globe today make ideal breeding grounds for novel forms of extremely virulent diseases. When combined with the typically irresponsible use of antibiotics, such diseases may also be expected to be immune to normal antibiotics (e.g., the resurgence of immune tuberculosis among the urban poor; the recent identification of a form of staph bacterium in Tokyo that is likewise immune to treatment). Finally, the influx of UV light from the depleted ozone layer can be expected to accelerate rates of mutation among all microorganisms, particularly viruses. (For more detail on all of the above, see McMichael 1993.)

Desertification

Warming implies a general movement toward the poles of the climates appropriate for the major food crops (wheat, rice, and maize). Unfortunately, at least in North America, the soils north of the (now global) "breadbasket" of the Great Plains are less fertile, implying a loss in yield with migration north. (Further, movement toward the poles implies increasing exposure to UV radiation.) For plants not under human cultivation, the polar shift in climate may outrun their ability to migrate, leading to their extinction (along with whatever other life forms depend upon them). The areas left behind are predicted to become prone to desertification. Indeed, portions of the U.S. state of Nebraska have rolling green hills that are actually sand dunes covered with grass. During the dust bowl era of the 1930s, the grasses died and the dunes moved with the wind as they reverted to desert. Current satellite data indicate that the same process has begun again. Pressures toward desertification are likely to grow, expanding deserts everywhere.

Severe Weather

The fundamental force powering all wind is heat, specifically the difference in heat between the equator and the poles. Hot, humid air rises high above the tropics and is blown toward the nearest pole. Along the way it cools, releasing its heat as rain. Storms are thus simply heat engines engaged in the impossible task of equalizing the temperature difference between the equator and the poles. The rotation of the Earth sets some of these storms to spinning like pinwheels, generating hurricanes (as they are called in the Atlantic) and cyclones or typhoons (their Asian equivalents). Sophisticated climate models run on supercomputers predict more frequent storms having higher windspeeds with increased warming, implying corresponding increases in deaths and infrastructural damage. In reality, the past two hurricane seasons in the Atlantic have been the most active of any on record over the past 20 years. Because the global population is growing, and most of that population lives near rivers and seacoasts, the mortality figures could become truly staggering as well as the cost of repair (already a demand on the budget of the United States and well outside the means of governments in the periphery). Adding to the difficulties of prediction here is that overall warming is still consistent with local cooling in certain areas for brief periods, leading to wild oscillations in annual temperatures and precipitation (Karl, Nichols, and Gregory 1997). More recent models have even suggested that the temperature difference between the poles and equator may decrease, implying a reduction in storm severity (Karl, Nichols, and Gregory 1997). Clearly atmospheric science has a need for much greater refinement ahead before it can reliably guide our expectations.

At the worst extreme, more far-fetched but still plausible results of sufficient warming could include a complete polar meltdown with sea levels shifted upwards of the full 250 feet. The consequent redistribution of mass from the poles (as ice) to the equator (as water) would potentially slow down the Earth's rate of spin, like an ice skater extending her arms while spinning. A change in the spin rate could alter plate tectonics even as the ocean cooled dramatically from the polar infusion. The former could stimulate major earthquakes and volcanic eruptions even as the latter caused hurricane force winds.

But setting aside these nightmare scenarios, even the changes observed so far have serious implications for the viability of mechanized agriculture and automated industry, each powered by the fossil fuels shown already to be so destructive to the biosphere.

THE GLOBAL REORGANIZATION OF PRODUCTION AND THE FISCAL CRISIS OF THE CORE STATES

Automation in the contemporary core is pervasive. Almost every act of consumption involves a machine at some stage, from cash registers to automated teller machines, while small computers are included in an ever-broader array of end-use commodities such as radios, ovens, and clocks. These examples are but echos of the real revolution experienced in production that began in the 1970s.

The hallmark of the relations of production characterizing capitalism is the structured antagonism between capital and labor. Precisely because capital needs labor to create commodities, it must solicit labor's cooperation and even support. Yet, of course, market competition continually encourages capital to reduce labor costs even while raising worker output. Labor, in its turn, can fight against these negative pressures from capital by slowdowns, strikes, and so forth. Until recently, the weapons available to capital to maintain its control were limited: the direct supervision of assembly, the importation of strikebreakers, and the employment of state-sponsored violence. Now, at last, during this century scientific investigation has yielded technologies that have enormously enhanced the power of capital against labor. These are, first, an infrastructure of telecommunications that allows for the remote control of multiple production sites and, second, the development of semi-intelligent machines that can potentially fulfill capital's ultimate dream of removing labor altogether. But an important cost of each of these developments has been an increase in the consumption of fossil fuels.

The motivation for the very first factories (in the eighteenth century) was not to increase the output per worker directly, but to do so indirectly by gathering all of the laborers together under the same supervised roof (Marglin 1974). That extension of the control of capital over labor expanded considerably in the late nineteenth and early twentieth centuries. In the steel industry, for example, one of the many methods created to divide laborers under the guise of "efficiency" was the imposition by management of artificial distinctions between workers of equal skill in the form of differing job titles and wages assigned to separate locations in the same assembly line (Stone 1974). The early decades of the twentieth century also saw Taylor's time-motion studies, which complemented and informed the perfection of the assembly line (Braverman 1974). Each of these successive reorganizations of production were explained in their day as technical improvements in efficiency, although the actual changes in output per worker were probably less dramatic than the extension of the political and social control of capital over labor.

In contrast with these earlier reorganizations of the shop floor, the successive and cumulative inventions of the vacuum tube (1920s), computer (1940s), and semiconductor (1950s) truly revolutionized production in a technical way. By endowing machines with the capacity to make "choices" (however crude), information technology has at last allowed machines to become genuine robot-workers, thereby enabling them to seize the bottom rung of the job ladder. In the United States, this has been the rung traditionally held by the unskilled or new immigrant (Aronowitz 1973), whose expulsion from the formal "monopoly sector" workforce of the core by the proliferation of robot technology has trapped them for the indefinite future in the informal sector. We can only expect that future technical change will eventually enable robots to climb ever-higher rungs of the job ladder, thereby expelling ever more "skilled" workers into the informal sector as their decision-making powers improve.

The revolutionary effects of the microchip started emerging in the form of robots on the shop floor in the 1970s and early 1980s, and were felt first by the working class. But by the mid-1980s the new technology had crept up to the levels of middle management, even as it spilled out of the most monopolized parts of the private sector into the offices of government. Specifically, the personal computer quickly transcended its initial role as a smart typewriter to encompass scheduling, accounting, statistics and scientific computation, blueprints for product design, movie animation and special effects, and so on. These multiple uses allowed for the combination of many different jobs onto one desktop, which in turn allowed for the elimination of an equivalent number of now redundant personnel and positions.

Insofar as these high-wage blue-collar and mid-range white-collar workers had been the core tax base of the welfare state, the dramatic contraction of their numbers during the microchip revolution (as manifested both by automation and relocation abroad of production facilities) eviscerated that base, exacerbating the growing fiscal crisis of the United States government at all levels as predicted and elaborated by James O'Connor as early as 1973 (O'Connor 1973). At that time he argued that the diminution of employment in the monopoly ("formal") sector was already threatening the supply of tax revenue, and history since has clearly vindicated his expectations: higher taxes on the shrinking number of high-wage workers in the monopoly and state sectors, combined with a contraction of employment in those sectors, and diminishing services available from the state have all coalesced to create an angry attitude toward the state, particularly within the ranks of the white males previously granted privileged entree into the monopoly and state sectors. By the late 1970s in the United States the implicit social contract (of a balance between wages, profits, and social services) among capital, labor unions, and the state that had reigned since World War II began to unravel. Starting in California with the tax revolt organized by Howard Jarvis in a popular referendum in 1978 and ratified nationally by the election of Ronald Reagan in 1980, the hostility of the downsized working classes and small business owners toward government and taxes has been skillfully manipulated by the right to cultivate antigovernment feelings strong enough to enthusiastically support the dismantling of the postwar welfare state.

Just as the United States led the technical transformation of agriculture among the countries of the core, so it has also led in the contraction of the welfare state among those same countries. Within the limits imposed by their respective histories of class struggle, Britain, Germany, the Netherlands, and even France have been compelled to follow suit during the 1980s and 1990s by constricting welfare outlays and tightening tax codes. In every case, these policies were aimed at coping with the same global problem: the collective evacuation of employment arising from automation and outsourcing within each of the local monopoly sectors. The global reorganization of production enabled by telecommunications and the microprocessor offered possibilities of higher profits from automation and factory relocation to the semiperiphery that were eagerly sought by monopoly firms in all of the nations of the core.

From an historically more detached perspective, the popularity of the revolt against the postwar welfare state can be understood as having been the political manifestation throughout the core of secular changes in technology and labor market structure apparent since the oil crisis of 1973 (Mandel 1978; Kotz, McDonough, and Reich 1994; McMichael 1996). Put another way, the social structure of accumulation prevailing since Bretton Woods—the social contract between capital and labor within the core and the financial institutions regulating relations between the core and periphery—fell apart during the 1970s. Real wages throughout the core have stayed at roughly the same level since the middle of that decade, unemployment has swelled until workers have been compelled to take jobs offering one-third of their former wages, and the resultant fiscal crisis stalled out the welfare state during the late 1970s. On the international level of global trade, in response to shrinking core markets, the debt crisis of the semiperiphery (whose export earnings depend entirely on core markets) became so bad that it compelled a serious reexamination of the viability of the World Bank Group in the early 1980s (McMichael 1996; Suter 1986). There were transient exceptions: the "Asian Tigers" of the semiperiphery were the beneficiaries of monopoly sector investment (the destinations for the flight of manufacturing and investment leaving the core). However, the rest of the semiperiphery and all of the periphery were less fortunate. Most of the former were saddled by debt, among whom some (e.g., Mexico and Brazil) precipitated the debt crisis of the early 1980s. Among the latter, Latin America and much of Asia struggled hard just to stay in place, while in Africa the economic situation actually deteriorated by every measure (see Terlouw 1992).

The collapse of the USSR and its Eastern European satellites can also be understood within this broader context of global disarray as a reflection of their secular digestion by the capitalist world-economy (Chase-Dunn 1982). Since World War II, a growing dependence on sales to the capitalist world for state revenue had gradually integrated the "socialist camp" into a role equivalent to the semiperiphery. For example, since the late 1970s Poland and Hungary had entered fully into the membership of the World Bank (Payer 1982). The collapse of 1989 can be understood as yet one more expression of the global depression and dis/reorganization afflicting the semiperiphery of the world-economy from 1973 to 1996 (the year of the official end of the "Uruguay Round" of the GATT and its conversion into the WTO).

Class Polarization and Ethnic Cleansing

What the planet's peoples have experienced over the last 20 years is the largest restructuring of the mechanisms of accumulation to have occurred since the Great Depression of the 1930s. The global mobility of capital combined with its command of armies of robots, both enabled by the microchip revolution, has finally broken the barriers to the equalization of wages between core and periphery by pitting their workforces directly against each other. Because the previous regime of unequal wages had sustained the high wages of the core and—via their taxes—the welfare state, the elimination of that core-periphery wage inequality has both threatened to eliminate that state and at the same time cast the lot of the workers of the core into the same pit as that of their brethren in the periphery. The long-term political results are as yet unpredictable.

In the periphery, the collapse of the patronage income derived from the alliance fears of the cold war (money spent on buying the allegiance of states in the periphery), combined with the tightening of loan restrictions from the World Bank, have reduced access to capital and blocked the ability of local ruling elites to compensate supporters with state contracts. In the semiperiphery (outside the favored group of "tigers" in Asia), the collapse of cold-war spending by the states of the core combined with limitations on the effective demand from markets in the core (again due to the contracting income of the working classes) have also reduced growth, which again has limited the ability of the governments to reward supporters. Finally, in the core the contraction of the welfare state has cut off benefits to the powerless even as it is rooted in the dropping lifestyle of the previously privileged. Throughout the globe, the gap between rich and poor has increased (World Bank 1992). One result has been the reemergence of the far right among the displaced White exiles from the monopoly sector in the core, along with the pervasive spread of nihilism among the urban young (whose nonpolitics of hopelessness, self-loathing, and body-piercing could go right, left, or simply erode into anarchy).

In many parts of the world, classes overlap with ethnic categories—occupations (hence incomes) have become historically and regionally associated with peoples sharing similar cultures and backgrounds. The United States abounds with examples: Mexican immigrants provide the bulk of labor for large corporate farms; Blacks and exiles from Central America provide the lowest wage unskilled work in hotels and restaurants along the east coast and in factories in the Midwest (Georgakas and Surkin 1975); motel chains have become increasingly owned by people from the Indian subcontinent; and many small shops in the ghettos of both coasts are now owned by Koreans (Bonacich 1994).

On a global scale and in historical perspective, social scientists have documented how certain ethnic groups who had already adapted to the role of small-scale retailers were deliberately encouraged and sometimes even relocated by colonial powers to be ethnic buffers between the general population and the rulers at the top (Bonacich 1994; Portes and Walton 1981; Wallerstein 1979). Specific examples of such middleman minorities are the Jews in Europe, the Indians in Africa and the Caribbean, the Chinese in Southeast Asia, and Korean shopkeepers in the U.S. today. Individuals from these national/ethnic backgrounds had already evolved into the niche of small-scale lenders and shopkeepers before European colonization. But their respective aptitudes were encouraged and facilitated by the colonial powers in the nineteenth century. With the help of these powers, Indians were settled in Africa and the northern coast of Latin America, Chinese were accelerated in their colonization of Malaysia and Indonesia, and the refugees from Jewish pogroms in Europe fled to New York.

Other racial and ethnic gradations have evolved quite independently of such outside "help." Latin and Central Americans assign low status to anyone defined as descended from the indigenous (preconquest) population, Europeans discriminate against southern Europeans in general and gypsies in particular, Russians loathe equally Jews and people from the south (e.g., Georgians, Kazaks); and Indians assume that those with darker skin are from the caste of untouchables.

This global overlap between class and race (always socially defined) exacerbates the political dangers of our era. In a time of generalized contraction of state expenditures, the falling living standards of those dependent upon those expenditures pits competing constituencies against each other. Insofar as these competing groups are often associated with different ethnicities, the political temptation to convert this competition into ethnic hatred is great. Within the countries of the core, the flight of capital, the defeat of labor, and the resultant fiscal crisis has fueled the rise of neofascism among the working class youth of France, Germany, England, and the United States. Their ideological response to the elimination of a saftey net is to attack "foreigners" with darker skins. Among the countries of Africa, the withdrawal of patronage income from the cold war has sometimes led to the collapse of the state altogether as ethnic peace could no longer be purchased with outside income (e.g., Ethiopia, Somalia, Rwanda, Burundi, Zaire, Congo, Liberia, and Sierra Leone).[14] Throughout the periphery other states face the same challenges, but may yet avoid similar disasters. Yugoslavia has demonstrated that even the semiperiphery is not immune. But even where total anarchic collapse has been avoided, major restructuring has been required as states scramble to catch up with the new mobility of capital (e.g., the bailout of Mexico following NAFTA and a similar IMF-sponsored rescue package for Thailand).

Put most simply, the entire international hierarchy of core, semiperiphery, and periphery prevailing since the earliest days of colonialism is being restructured by the new mobility of capital enabled by the microchip, leading to a conversion of the stratification between nations into a stratification within nations. Accordingly, we are living in a time when the traditional hostility between nations is being replaced by a class/ethnic hostility within them. Global war is being augmented by civil war that is ethnically based. All of this restructuring is made possible by the microchip, which is in turn dependent upon an industrial base fueled by fossil fuels, which in their turn power global warming, which in its turn must ultimately contract global food supply and drive food prices up. Unless global surplus value is redistributed into higher wages to support higher food costs, the necessary result must be sharper conflicts and greater killing.

Inter-Core War

One last factor left hitherto untouched is the prospect of global war. This topic is sufficiently complex that it has generated a substantial literature (e.g., Chase-Dunn and Bornschier 1998; Goldstein 1988; Thompson 1990; Modelski and Thompson 1988; 1998). The central question motivating many of these works is whether global war is cyclical and, if so, what the mechanisms governing its frequency are. The answer adopted by many of these authors is that it is indeed cyclical, that the governing cycle has a period of between 40 and 60 years, and that the recurrent issue is the need to update the interstate hierarchy implicit within the existing international political structures to reflect the economic changes in that hierarchy experienced since the preceding inter-core ("world") war. Put more simply, world wars are the means by which the obsolete global political institutions are smashed and reorganized to make them reflect the new economic power hierarchy. Should the ordinary war cycle continue, the likelihood for the next war should be greatest around 2010 to 2020. If, however, one factors in the current apparent collapse of the states in the periphery, the growth of inequality in the core leading to worsening ethnic tension and state delegitimization as disparities within the core reflect ever more faithfully the disparities outside of it, and, finally, the uncertainties of harsher weather and rising food prices addressed here, then the resultant threats to global security (however defined and by whom) may greatly accelerate the timing of the war cycle. Finally, should such a war break out, it will almost certainly be nuclear (if only in part), in which case every single biospheric problem cited above will be grossly worsened and the probability of total social collapse made far more likely.

CONCLUSION

The title of this essay is a deliberately mixed metaphor. The "Killing Fields" refers to the genocide perpetrated on the people of Cambodia by the administration of Pol Pot. The "Horsemen" refer to the legendary four forces of the "Apocalypse" found in the book of Revelations in the Bible: starvation, disease, pestilence (insects), and war. The former was, in retrospect, an early example of the genocidal collapse of a peripheral state, the latter a plausible prediction about our collective future based on biospheric and world-system processes. My choice of this title was to suggest the linkage between the capitalist imperative to use fossil fuel technology against labor and the destruction of the biosphere; how that ecological destruction will necessarily entrain starvation, disease, and war; and, finally, how the stresses created by this social dissolution will enhance the likelihood of wars of "ethnic cleansing."

Unchecked global capitalism has put us in a double-bind: if the supply of fossil fuels were unlimited, global warming would eventually eliminate arable land and cook us to death. Alternatively, assuming the supply of fossil fuels is limited, their exhaustion will drive up prices for food, housing, and transport to levels beyond the reach of most people, fueling class and ethnic conflict. Either scenario is fatal. But the second is more likely. Although stocks of fossil fuels unknown today may yet be discovered, the energy required to locate and retrieve them must eventually grow exponentially.[15] This fact will inevitably manifest itself as increasing energy prices. Obviously alternative energy sources will become more frequent and perhaps even dominant. However neither solar, geothermal, nor nuclear energy can be converted into fertilizers, herbicides, and pesticides. These attributes are unique to fossil fuels. As they diminish, agricultural yield per acre must fall and food prices rise proportionally. This is particularly true of the soils in the tropics, which have by now become almost completely dependent on imported chemicals based on fossil fuels (McMichael 1996; Colinvaux 1978).

When one includes in this picture the probability that robots could displace all but the most highly trained "knowledge workers," the scenario becomes bleak: a huge mass of malnourished and desperate people struggling to live from a shrinking black market trafficking in illegal goods, packed tightly together in dilapidated ghettos rife with virulent diseases and gang violence, and sporadically involved in ethnic/neighborhood/gang wars. Even in the absence of another inter-core and nuclear war, the extremely high death rates that would accompany this grim scenario would, over several generations, reduce the global human population back down to levels supportable by low-energy agriculture—perhaps the two to three billion we numbered globally before World War II. Of course nuclear war would itself dramatically reduce population levels as well as infrastructural support. But even today, we must remember that the destruction of the tropics is creating a mass extinction already worse than that accompanying the loss of the dinosaurs (Wilson 1990, 1992). For example, today there remain no more than 3,000 tigers alive on the entire planet (National Public Radio, *All Things Considered,* July 9, 1997). The numbers of black rhinoceros are similar and lions not far behind. But for every charismatic large animal endangered, there are millions of species of smaller insects, plants, and amphibians disappearing with every acre of deforestation, not just in the tropics but as well across the countries of the core. To the extent that these issues have been publicly addressed at all, the problem has been laid at the doorstep of population growth particularly in the periphery. But it should by now be clear that such a belief is at best misinformed: both the growth in population and the threats to life on the planet in general are caused by the use of technology solely for the pursuit of capital accumulation on a world scale.

The human historical record reveals many examples of the rise of societies whose complex organization allowed for the production and distribution of enough food that all could live. But the same record also shows that every one of these societies crumbled, often because of resource depletion accompanied by civil and/ or border wars (Tainter 1988). The argument that I have presented here merely shows that we continue to be constrained, as were our ancestors, by the boundaries imposed by the physical laws of energy flow (thermodynamics). Unlike them, we have the intellectual tools to predict and even avoid our fate. But these tools and knowledge are ignored and unheard when they imply social changes that are contrary to the needs of those in power.

The power for change lies with the general population worldwide. If they remain uninformed, their spontaneous revolts will simply accelerate the descent into anarchy. But if properly informed, the popular mobilization required to reorganize the social order—already implicit within the delegitimation of the current states—can be focused into a coherent global objective. The occurrence of several major natural disasters may yet be necessary to provide the correct unifying vision. Tragic as this may be, to paraphrase Samuel Johnson, "There is nothing like a hanging to concentrate the mind."

NOTES

1. At its most simplistic level, this collision appears to be merely a restatement of Malthus: the human population is growing faster than our ability to increase food yield per acre. However there is an important modification to this apparent acceptance of Malthus that must also be included. Malthus and his followers have tended to blame poverty on "over-population," thereby shifting the blame for income inequality away from capitalist accumulation to the victims of that accumulation. In stark contrast to this temptation to "blame the victim," the analysis here will demonstrate that the primary culprits of our collective predicament are to be found in the mechanisms that sustain the rich, not in those that perpetuate the poor.

2. The lingering presence of conifers in the Smokey Mountains of southern North America provides silent testimony to the southern extent of the last glaciation. Once the dominant species, they have eventually been pushed out by the warmth driving the glacial retreat, yielding their territory to the better equipped deciduous. Those conifers that survived did so by gradually climbing higher up the mountainsides, where they remain today stranded thousands of miles south of their brethren in the arctic.

3. The red soils of the southern United States seem at first an exception to this generalization, because they are fertile when cared for. However, it may be that the lower incidence of solar energy there as compared to the tropics has prevented the development of a biomass as aggressively efficient in extracting food as that found in the tropics.

4. Exactly consistent with the Darwinian "exclusion" principle explained by Bonner (1988).

5. The two most obvious exceptions are the Mayan and Kmer (Angor Wat) civilizations, each relatively recent entries to the historical stage. However each was located on floodplains, whose local fertility is greatly enhanced by the importation of runoff from all of the catchment areas upstream.

6. An attribute the ecologist Colinvaux nicely captures by the label of "niche-shifting" (1978, chap. 16).

7. The label "tributary" was originally applied by Amin (1976) to characterize the principal form of surplus accumulation used by the pre-capitalist empires. It refers to the institutional forms that accumulation took: on an individual level, the peasant or slave that worked the land was expected to give up a portion of his crop to the representatives of the state as a "tribute" or tax; while surrounding client or colonial states purchased their continued nominal "independence" also by the payment of "tribute." On both levels, the revenue of the state was a form of what would now be called a "protection racket." For further information see Amin (1976) or Chase-Dunn and Grimes (1995).

8. Even though it did greatly catalyze the conversion of large areas of land from traditional uses to sheep production while simultaneously encouraging the foundation of plantations using slave labor for cash-crop production.

9. An excellent account of the development of these higher incomes in the colonies of European settlement can be found in Amin (1976).

10. The regimentation of worker movements in commodity assembly based upon the time-motion studies of Frederick Taylor, extensive discussion of which can be found in Braverman (1974).

11. Fossil fuels are solar in origin as well, but insofar as their initial solar "charge" happened during and before the dinosaur eras, our use of them now is like taking money out of a savings account without any restoration. We are living off of our solar "capital."

12. A longer and more specific list can be found in Commoner (1971).

13. Dengue fever has already been observed moving north from Central America into Mexico.

14. Ironically counter to this trend, the withdrawal of cold war patronage has actually facilitated peace agreements and democratization in El Salvador and Guatemala because the support of the United States was essential to the power of the military supporting the oligarchies there.

15. This is another manifestation of the second law of thermodynamics elaborately and persuasively explained by Commoner (1977).

REFERENCES

Amin, Samir. 1976. *Unequal Development*. New York: Monthly Review Press.

Aronowitz, Stanley. 1973. *False Promises: The Shaping of American Working Class Consciousness*. New York: McGraw-Hill.

BBC. 1997. "Outlook." May 14.

BBC. 1997. "The World Today." May 8 and May 19.

Blaustein, Andrew R., and David B. Wake. 1995. The Puzzle of Declining Amphibian Populations. *Scientific American* 27(4): 52-57.

Bonacich, Edna. 1994. *The New Asian Immigration in Los Angeles and Global Restructuring*. Philadelphia: Temple University Press.

Bonner, John Tyler. 1988. *The Evolution of Complexity by Means of Natural Selection*. Princeton: Princeton University Press.

Braverman, Harry. 1974. *Labor and Monopoly Capital: The Degradation of Work in the Twentieth Century*. New York: Monthly Review Press.

Burbach, Roger, and Patricia Flynn. 1980. *Agribusiness in the Americas*. New York: Monthly Review Press and the North American Congress on Latin America.

Caldwell, Lynton Keith. 1996. *International Environmental Policy*, 3rd ed. Durham, NC: Duke University Press.

Chase-Dunn, Christopher, and Peter Grimes. 1995. World-System Analysis. *Annual Review of Sociology* 21: 387-417.

Chase-Dunn, Christopher, and Tom Hall. 1997. *Rise and Demise: Comparing World-Systems*. Boulder, CO: Westview Press.

Chase-Dunn, Christopher, and Volker Bornschier, eds. 1998. *The Future of Hegemonic Rivalry*. London: Sage.

Chase-Dunn, Christopher. 1982. *Socialist States in the World- System*. Beverly Hills, CA: Sage.

Chew, Sing. 1996. "Accumulation, Environmental Degradation, and Core-Periphery Relations in the World-System, 2500 BC to 1990 AD." Paper presented at the 36th annual meetings of the International Studies Association, Chicago, Illinois, February 21–25.

Colinvaux, Paul. 1978. *Why Big Fierce Animals Are Rare*. Princeton, NJ: Princeton University Press.

Commoner, Barry. 1971. *The Closing Circle: Nature, Man, and Technology*. New York: Bantam.

Commoner, Barry. 1977. *The Poverty of Power: Energy and the Economic Crisis*. New York: Bantam.

Folbre, Nancy. 1977. Population Growth and Capitalist Development in Zongolica, Veracruz. *Latin American Perspectives* 4(4): 41-55.

Georgakas, Dan, and Marvin Surkin. 1975. *Detroit: I Do Mind Dying. A Study in Urban Revolution*. New York: St. Martin's Press.

Georgescu-Rogen, Nicholas. 1971. *The Entropy Law and the Economic Process*. Cambridge: Harvard University Press.

Gibbons, John H., Peter D. Blair, and Holly Gwin. 1990. Strategies for Energy Use. In *Managing Planet Earth: Readings from Scientific American Magazine,* 85–96. San Francisco: Freeman.

Goldstein, Joshua. 1988. *Long Cycles: Prosperity and War in the Modern Age*. New Haven, CT: Yale University Press.

Grimes, Peter. 1982. Poverty, Exploitation, and Population Growth: Marxist and Malthusian Views on the Political Economy of Childbearing in the Third World. Master's thesis, Michigan State University.

IPCC (Intergovernmental Panel on Climate Change). 1990. *Climate Change: The IPCC Scientific Assessment*, ed. J. T. Houghton, G. J. Jenkins, and J. J. Ephraums. New York: Cambridge University Press.

IPCC (Intergovernmental Panel on Climate Change). 1992. *Climate Change 1992: The Supplementary Report to the IPCC Scientific Assessment*, ed. J. T. Houghton, B. A. Callander, and S. K. Varney. New York: Cambridge University Press.

Karl, Thomas R., Neville Nichols, and Jonathan Gregory. 1997. The Coming Climate. *Scientific American* (May): 79–83.

Kotz, David M., Terrence McDonough, and Michael Reich. 1994. *Social Structures of Accumulation*. New York: Cambridge University Press.

la Riviere, Maurits. 1990. Threats to the World's Water. In *Managing Planet Earth: Readings from Scientific American Magazine,* 37–49. San Francisco: Freeman.

Maddison, Angus. 1995. *Monitoring the World-Economy 1820–1992*. Paris: Organization for Cooperation and Development in Europe, Development Centre Studies.

Mamdani, Mahmood. 1972. *The Myth of Population Control: Family, Caste, and Class in an Indian Village*. New York: Monthly Review Press.

Mandel, Ernest. 1978. *The Second Slump*. London: New Left Books.

Marglin, Stephan A. 1974. What do Bosses Do? The Origins and Functions of Hierarchy in Capitalist Production. *Review of Radical Political Economics* 6(2): 33–60.

McMichael, A. J. 1993. *Planetary Overload: Global Environmental Change and the Health of the Human Species*. New York: Cambridge University Press.

McMichael, Philip. 1996. *Development and Social Change: A Global Perspective*. Thousand Oaks, CA: Pine Forge Press.

Mitchell, Brian R. 1993. *International Historical Statistics: The Americas 1750–1988*, 2nd ed. New York: Stockton Press.

Modelski, George, and William R. Thompson. 1988. *Seapower in Global Politics, 1494–1993*. Seattle: University of Washington Press.

Modelski, George, and William R. Thompson. 1998. *Innovation, Growth and War: The Co-Evolution of Global Politics and Economics*. Columbia, SC: University of South Carolina Press.

National Public Radio. 1997. *All Things Considered.* Broadcasts on May 9, June 5, July 9.

O'Connor, James. 1973. *The Fiscal Crisis of the State*. New York: St. Martin's Press.

Painter, Kim. 1997. Puberty Signs Evident in 7- and 8-Year-Old Girls. *USA TODAY,* April 8, 1.

Payer, Cheryl. 1982. *The World Bank: A Critical Analysis*. New York: Monthly Review Press.

Perlman, Michael. 1977. *Farming for Profit in a Hungry World: Capital and the Crisis in Agriculture*. New York: Universe Books.

Peters, Robert L., and Thomas E. Lovejoy. 1990. *Global Warming and Biological Diversity*. New Haven, CT: Yale University Press.

Pimentel, David, and C. Harvey. 1995. Environmental and Economic Costs of Soil Erosion and Conservation Benefits. *Science* 267(Feb. 24): 1117–23.

Portes, Alejandro, and John Walton. 1981. *Labor, Class, and the International System.* New York: Academic Press.

Rudel, Tom, and Chun Fu. 1996. A requiem for the Southern Regionalists: Reforestation in the South and the Uses of Regional Social Science. *Social Science Quarterly* 77 (Dec.): 804–20.

Runnels, Curtis N. 1995. Environmental Degradation in Ancient Greece. *Scientific American* 272(3): 96–99.

Sanderson, Steven K. 1995. *Social Transformations: A General Theory of Historical Development*. London: Basil Blackwell.

Stevens, William K. 1997. Five Years After Environmental Summit in Rio, Little Progress. *New York Times,* June 17, C-8.

Stevens, William K. 1997. Something to Sing About: Songbirds Aren't in Decline. *New York Times,* June 10, C-1.

Stone, Katherine. 1974. The Origins of Job Structures in the Steel Industry. *The Review of Radical Political Economics* 6(2): 61–97.

Suter, Christian. 1986. *Debt Cycles in the World-Economy*. Boulder, CO: Westview.

Tainter, Joseph A. 1988. *The Collapse of Complex Societies*. New York: Cambridge University Press.

Terlouw, Cornelius Peter. 1992. *The Regional Geography of the World-System: External Arena, Periphery, Semiperiphery*. CoreUtrecht: Faculteit Ruimtelijke Wetenschappen, Rejksuniversiteit Utrecht.

Thompson, William R. 1990. Long Waves, Technological Innovation and Relative Decline. *International Organization* 44: 201–33.

U.S. Bureau of the Census. Washington, DC: United States Government Printing Office.

Wallerstein, Immanuel. 1979. *The Capitalist World-Economy*. New York: Cambridge University Press.

Wilson, Edward O. 1990. Threats to Biodiversity. In *Managing Planet Earth: Readings from Scientific American Magazine,* 49–60. San Francisco: Freeman.

Wilson, Edward O. 1992. *The Diversity of Life*. New York: Norton.

World Bank. 1982. *World Tables.* Washington, DC: World Bank Group.

World Bank. 1995a. *World Development Report*. Washington, DC: World Bank Group.

World Bank. 1995b. *World Tables*. Washington, DC: World Bank Group.

3

Ecosociology and Toxic Emissions

Albert J. Bergesen & Laura Parisi

INTRODUCTION

Environmental studies has brought a new set of questions to social theory in general and world-system theory in particular. Put in the simple terms of social research, environmental concerns constitute both an independent and dependent variable for world-system theory. As an independent variable, radical environmental ethics question the moral width of theories of social transformation and their continued privileging of human concerns at the expense of nature (Bergesen 1995a, 1995b). On the other side of the equation, the environment is a massive dependent variable providing a vast array of indicators of environmental degradation that can be plugged into explanatory world-system models allowing us to estimate to what degree environmental outcomes are products of world-system processes. This chapter concerns the green challenge to world-system theory as both independent and dependent variable. Concerning the moral challenge, we introduce two conceptions of environmental justice: justice for humans and justice for the environment. This distinction corresponds to the Social Ecology/Environmental Justice versus Deep Ecology positions within the environmental movement. It is one of our purposes here to try to bridge some of the differences between these traditions. After a discussion of these issues, we turn to a quantitative study linking dependent development with rates of toxic emissions.

To begin with, world-system theory and environmental studies are a natural fit. It is surprising, therefore, that they aren't more interconnected, as they share a common global holism. While there are local environmental issues in virtually every region of the world, the overarching environmental assumption is of the global character of the ecosystem. Acid rain may initiate in the production process of one country, but it can fall on the trees of any number of countries, and a growing hole in the ozone layer affects living things around the world. A similar global holism is found in world-system theory, where authors may vary in emphasizing political, economic, or cultural aspects of the global totality, yet agree that what is desired is a social science that deals with the distinctly global level of systematic human interactions.

ENVIRONMENTAL JUSTICE

World-system theory has been concerned with the justice and equity of the global structuring of human affairs as the material conditions of the peoples on the edges, or periphery, of the world-system are linked to the needs, power, and material advantages of those who constitute its center, or core. The study of environmental justice presents a new opportunity to employ world-system theory. To begin with there are two interconnected, yet separable, interpretations of environmental justice. The question is justice for whom? Industrial pollution in the neighborhoods of workers and the poor is a question of justice for humans, and on a world scale a similar issue might be the relocation of polluting plants from the North to the South, where profits go North and hazardous environmental wastes remain South.

A second meaning of environmental justice centers upon justice for the environment, the notion that environmental degradation has significance for the environment itself, independent of what it means for humans (the Deep Ecology idea). Here the issue of justice centers upon the rights of nature to live unexploited by human need, greed, or development. The problem is environmental and human priorities confront each other. From a humanistically progressive perspective, economic development is a valued end, and world-system theory helps understand how the development of some parts of the world takes place at the expense of others. But such human advancement can be at the expense of nature. Humans are obviously a part of nature and separating them out makes no sense, yet we are aware of the differences that arise from an environmentalism that focuses upon environmental degradation from the point of view of humans as opposed to the point of view of the environment.

For example, from the human justice point of view equalizing toxic risks between rich and poor countries, or making rich countries pay more for the environmental damage their corporations create in developing countries, constitutes human justice. But from the point of view of justice for the environment—for nature as a living thing—that humans are equally at risk, or pay their fair share, or that the producer of pollutants cleans up, says nothing. From an ecocentric point of view, making things morally right within one species (humans) says nothing per se about relations between species or about nature as a whole. Abundance and justice for all humanity may very well be at the expense of the rest of nature, as it has often been argued that if the world lived at the level of the developed countries the ecosystem would be overloaded and exhausted. Even if overall living standards were lower, the point remains that the equalization of rights, opportunities, and well-being among humans says nothing about relations between humans and other species. Human justice (justice within a species), then, does not of logical necessity entail ecojustice (justice between species and among all living things).

Some social ecology positions argue that attaining human equality would put humans on a different relationship with their environment. From the perspective of Eco-Marxism the transformation of the human means of production would end environmental degradation, and Eco-Feminism suggests that the transformation of gender relations will in turn transform human relations with nature. These are logical positions at first glance. If it is the activities of one species that cause damage to nature, it makes perfect sense to theorize that restructuring these relations should lead to a different relationship with the environment.

But the issue may be more complex. Is it these human social formations (capitalism, patriarchy, racism, etc.) that result in the detrimental human-to-nature relation, or is it a detrimental human-to-nature relation that results in capitalism, racism, ageism, or patriarchy? Now no one wants to blend the social into the biological for the political reason of removing the possibility of willful social change from the human agenda. On that we all agree. Yet the assumption of received social theory that domination, exploitation, and subordination autonomously arise within the human species when one class, gender, race, or age group takes advantage of another, creating, over time, capitalism, patriarchy, racism, and other structures of domination, is based on the false assumption of the autonomy of human social formations. This assumption is susceptible to the very critique social theory leveled against the false individualism of utilitarianism and classical economic theory in arguing that individual choice is not aggregated to comprise institutions, but institutions shape individual choice. Paraphrasing Marx, it is not the individual buying and selling that makes capitalism, but capitalism that makes the individual buy and sell. Social formations were the wider web within which human choice resided, hence they seemed, if not determinate, then at least a constraint that had to be factored in when explaining individual decisions. From the perspective of social theory, the individual is the part and the social formation the whole. But from an ecocentric perspective, the social formation is but a part and the larger ecological web the whole. What made social theory radical was its claim to represent the totality of determinate relations, the context within which the classical economist's or today's rational choice theorist's self-propelled and self-motivated individual resided. But yesterday's social totality is today's part, in an ecosociology of even larger, more encompassing species to species relationships.

ECOSOCIOLOGY

Is it relations within humankind that determine human with nature relations, or is it human relations with nature that determine the within species relations that comprise social formations? The history of social thought has been to always raise the level of analysis to an ever wider web, and now we may have reached the point where expanding that web means moving social theory to a species to species level. When species to species relations become the definition of social structure on the eco level of analysis, then all human social formations are but one of many actors. The most powerful actor, yes, but because the capitalist was powerful did not eliminate the fact that this power arose from the larger web that was the capitalist mode of production. The question we are trying to raise here is to what extent is human power a product of the human structural location in something like an ecosocial system? We understand the importance of brain, language, thumb, upright posture, and other innate attributes of humans. What we wonder is whether such essentialism is the only way of understanding the negative consequences humans have for the environment.

In effect, if we were to move toward something like an ecosociology, could we then perform an ecostructural analysis the same way we presently perform social-structural analysis, where the human species could then be considered something like an eco-class actor within a larger ecosocial system? That is, can we ratchet up the notion of structure from its present intra-species specification and make it applicable to inter-species relations without dissolving ecosociology into biology? That is the question, and at present there is no obvious answer.

We would like to suggest that this is not only a possibility, but a necessity, as the moral rights of nonhuman beings have to be brought within theories of not only moral/political philosophy but also within theories of the structuration of living things—that is sociology—except now, with nonhumans included, with what will have to become ecosociology. In such an ecosociological analysis, present theory, which emphasizes the auto-transformation of human social forms without regard to their ecosociological location, is just as conservative and naive as earlier rational choice theory, which emphasized human choice without regard to the web of human relations within which it is embedded. From an ecosociological point of view, then, radical theories of social change are not so radical, for they assume human self-transformative possibilities without regard to human ecostructural location. Received social theory, then, is something of a self-serving humanocentric ideology, postulating a mythical autonomy to human action without regard to human ecosociological position. In effect, if individual self-transformation cannot be granted without factoring in the social structures within which it is embedded, human social transformations also cannot be autonomous of the ecosociological web within which they are nested. We are engaged in collective Robinson Crusoeism if we continue to posit the autonomy of the social from the laws of motion of the ecosociological.

On an ecosociological level, then, radical social theory is nothing more than rational choice theory for humanity as a whole, a species ideology about species self-transformation without regard to the larger set of ecosociological relations within which that species is embedded. The deeper point here is that the human social totality is no longer the relevant totality for the theoretical question of embeddedness and hence no longer relevant for the question of structural determination. Radical transformations of structures, as program and theory, have to be ratcheted up a notch such that the human project is correctly seen in its embedded position within the larger, wider web of ecosociological relations. Radical transformative projects, then, have to deal with the widest—and hence most determinative—web of structural relations which are no longer social relations among humans, but ecosociological relations among all living things.

The implications for world-system theory are clear. It had once been seen as the largest and widest web of association and determinate structure; hence it was the object to study and the structure to transform. But from an ecosociological position, the world-system is only the structural web within one of many species, a substructure within a larger ecopattern. This is not to argue for any sort of biological determination. The point is not to naturalize the social but to socialize the natural. What is proposed here is an awareness that structuration does not end with human-to-human relations—even if they are worldwide. The challenge is to widen the moral and structural web and thereby relocate world-systems within ecosystems and theorize how ecosystemic functioning affects world-system functioning.[1]

Here, at the end of the twentieth century, the firm ontological divide between humans and other living things is dissolving. Scientific research is breaking down the humano-essentialism which undergirds classical political thought. The discussion seems most developed concerning questions of the moral and political rights of nature. This, though, is but the beginning. What is next is the question of how wide is the web of social structuring. Up to now, it has stopped at the boundary of the human species, ideologically justified by increasingly untenable hypotheses postulating essential differences between humans and animals. From Adam Smith and Karl Marx to Weber and Durkheim and the social constructionists in today's environmental movement, what is natural has been seen as either apart from what is human or as a construction of the human imagination. But this position will not stand. Social forms will be retheorized as ecoforms and the web of determinate relations between living things will be widened from its present imprisonment within the borders of humanity's speciesdom. The social as the maximal sense of the collective and of the totality will be replaced by the ecosocial, as we are in the midst of the next great transition in theory where social relations will be replaced by ecorelations, social formations by ecoformations, and social justice by ecojustice.

PREDICTING TOXIC EMISSIONS

We turn now from the more general critique of sociological analysis to a more specific study: predicting world-systemic affects upon cross-national emissions of toxic materials. Such a study of the environment from the point of view of the world-system is itself a relatively new area of research (Smith 1994). Work has begun on deforestation (Burns et al. 1994; Chew 1996; Kick et al. 1996; Rudel 1989); raw material extraction (Bunker 1984, 1985, 1994; Bunker and Ciccantell 1995); carbon dioxide (CO_2) production (Dietz and Rosa 1997; Roberts and Grimes 1997); toxic emissions (Hettige, Lucas, and Wheeler 1992; Low and Yeats 1992; Lucas, Wheeler, and Hettige 1992); and participation in environmental treaties (Roberts 1996). In the research reported here, we will extend the world-system perspective to explaining international rates of toxic emissions into the environment. In this preliminary analysis we will test hypotheses linking political and economic development along with economic dependency to rates of toxic emissions. As far as we know there has been no previous application of dependency theory to explain differential rates of toxic emission within the capitalist world economy.

Independent Variables

Our indicator of political development is the presence of democratic practices in the political system. We use a ten point scale based on eight items: openness, competitiveness, and regulation of executive recruitment, monocratism, constraints on the chief executive, regulation and competitiveness of political participation, and centralization of state authority (Jaggers and Gurr 1996). Our measure of economic development is Gross Domestic Product per capita (Summers and Heston 1991), and our indicators of economic dependence are percent of exports in a single commodity (UNCTAD 1983, 1984, 1986, 1988, 1989, 1990) and direct, foreign private investment (IMF 1991, 1995).

Dependent Variable

Our indicator of toxic emission is based on the U.S. Environmental Protection Agency's (EPA) Toxic Release Inventory (TRI), which used a sample of 15,000 industrial plants to record air, water, underground, and solid waste releases of some 320 toxic substances from each reporting plant. From this they obtained toxic emission rates for different industries. While the TRI has been pressed into service in a number of studies, this indicator has some shortcomings. There are no doubt more than the 320 toxic substances identified by the EPA in the U.S. economy; only a small portion of the facilities that might use such substances are part of the TRI; and the data is self-reported and noncompliance may be high (Szasz and Meuser 1997, 8). "Still, TRI was and is a major step forward. For the first time, researchers had comprehensive, nation-wide data, however problematic, on toxic emissions from operating plants" (Szasz and Meuser 1997, 9). With this data researchers examined toxic emissions near Los Angeles (Szasz et al. 1993; Burke 1993), across Florida (Pollock and Vittas 1995), and around Pittsburgh (Glickman and Hersh 1995). In the research reported here we rely upon a measure of total toxic emissions for a sample of 70 countries that is based on TRI data (Lucas, Wheeler, and Hettige 1992). This data set was comprised by taking output data from the U.S. 1987 Census of Manufactures for each of these 15,000 EPA plants to then calculate the aggregate toxic releases per unit of output for each of the 37 ISIC industrial categories on which the EPA data on toxic emissions was gathered. From this a Total Release Intensity score was created comprised of the total pounds of all 320 toxic releases (whether atmospheric, solid or effluent) per dollar's worth of output. Then United Nations annual sectoral data for a sample of countries was multiplied through by the U.S. intensity rate for each of the 37 industrial sectors to give each country in the sample a total toxic emission score.

As with the TRI there are some questions about this technique. First, it assumes an equality of industrial production procedures across zones of the world-system. For some industries this may be problematical as age of plant, level of technology, quality of labor, and environmental standards no doubt differ significantly in developed and underdeveloped economies. For their cross-national comparisons, Lucas, Wheeler, and Hettige (1992, 72) claim that "most existing empirical work assumes rough constancy in relative cross-sectoral pollution intensity, invariably identifying the same sets of 'heavy polluters' (e.g., metals, cement, pulp and paper, chemicals) and 'light polluters' (e.g., most light assembly, food products, instruments)." They claim this assumption is based on reports of investments on pollution control equipment and case studies done by regulated industries in OECD economies since 1975. The total toxic emission score, then, is a preliminary measure of toxic emissions. It represents the only way to estimate emission rates for countries around the world. Hence its use as a preliminary benchmark measure.

Economic Development

Our first analysis centers upon the association between economic development and toxic emissions. This relationship has been hypothesized to be curvilinear. Hettige, Lucas, and Wheeler (1992, 479) have spoken of "three stages of industrial development dominated by: (1) agroprocessing and light assembly, which are (relatively) low in toxic intensity; (2) heavy industry (e.g., metals, chemicals, paper), which has high toxic intensity; and (3) high-technology industry (e.g., microelectronics, pharmaceutical), which is again lower in toxic intensity." Changes in pollution rates are theorized as a natural shift from mode of production to mode of production in the evolution of economic life, with this being "in part . . . perceived as a natural evolution" (Hettige, Lucas, and Wheeler 1992, 479). (There is no dependency or world-system perspective here, where the development of one zone [the core, the North, the developed, rich countries] differentially affects the development of others [the periphery, the South, the underdeveloped, poor countries]). Underdevelopment is not assumed. Development is theorized as linear and all countries are assumed to pass through roughly the same stages. What is interesting here is that this curvilinear relationship seems to apply to a variety of pollutants. This has led some to speak of something like an "environmental Kuznets inverted U-curve" for pollutants, such as CO_2 (Grimes and Timmons 1995) and a variety of toxic wastes (Hettige, Lucas, and Wheeler 1992; Lucas, Wheeler, and Hettige 1992).

Their reasoning is as follows. From the broadest evolutionary perspective, hunter-gatherer societies are seen as more efficient, using less, and existing closer to the land, hence emitting less pollutants in the economic process of reproducing their existence. This has also become the moral position of many green philosophers, bioregionalists, and in general the small-is-better outlook of much green morality. As evolution proceeds and the means of production develop, emissions increase, reaching a high point during the stage of heavy industry where per some unit of production the output of pollutants would be highest and would then level off in late industrialization where less polluting industrial processes, greater government environmental controls, and cleaner industry (service sector growth) yield a lower pollution rate. In this regard it has been estimated that recent lowered energy intensity in the United States is about 45 percent attributable to shifts in the economy toward the production of less energy-intensive materials, as the production of steel and cement, for instance, has fallen and that of electronic goods has dramatically risen. The other 55 percent has been attributed to more efficient production processes and equipment. Some of this seems tied to the efficiency measures brought on by the higher competition of the B-phase downturn as firms strive to maintain profit levels in a shrinking and reorganizing world economy (Flavin and Durning 1988, 53).

While the curvilinear pollution hypothesis is interesting, is there any evidence for this proposed inverted U-shaped relationship between rates of pollution and levels of development? Two studies report such a relationship. For Grimes and Timmons (1995, 1) the idea of an environmental Kuznets curve refers to "whether effluents produced by industrial processes increase monotonically or curvilinearly with development." By looking at the relationship between CO_2 emissions per unit of GDP and GDP per capita as a measure of economic development, they report a curvilinear relationship between carbon dioxide emissions and development. This curvilinear relationship was also reported for toxic emissions by Lucas, Wheeler, and Hettige (1992, 69), who find a curvilinear relationship between toxic emissions per unit of GDP and GDP per capita as a measure of development. We replicate this analysis. Using the toxic estimates for a sample of countries (see appendix) we plot log toxic emissions divided by total GDP by GDP per capita squared. We found toxic emission goes up with a rise in GDP/pc, but at a certain point begins to level off, supporting the hypothesis that the mix of cleaner industry and the growing service sector of advanced capitalist countries results in lower rates of toxic emissions.

What is to be made of this curvilinear finding? There are at least two possible interpretations. First, it may be that more GDP is extracted per unit of pollutant emitted in high-income countries, as they are more efficient. They have more high-quality and newer equipment, more advanced techniques to do the same thing, and stronger environmental laws. Second, it may be that the growing service sector of high-income countries means less manufacturing is occurring per unit of GDP, hence less production opportunities to pollute per unit of GDP. Here level of efficiency is not the issue; all countries may pollute at the same rate—an industry is an industry argument. It is rather that core countries have a smaller proportion of their total economic activity involved in basic industries that emit toxic pollutants. In this regard, Lucas, Wheeler, and Hettige (1992) found that toxicity per unit of manufacturing (as opposed to per unit of GDP) did not decline for high-income countries and concluded that the earlier finding reflected the declining proportion of manufacturing within the total GDP of high-income countries. "When intensity [and toxic emissions] is defined as industrial emissions divided by GDP, the rise in intensity with respect to rising income at least tapers off at higher income levels. On the other hand, when intensity is defined as pollutants per unit of manufacturing output this is definitely not the case (indeed, the relationship . . . goes the other way)" (Lucas, Wheeler, and Hettige 1992, 76). A similar conclusion was reached for CO_2. While high-income economies appear to be more efficient, the overall amount of carbon dioxide is still increasing. "In raw volume, both carbon dioxide emissions and energy consumption are closely correlated with the size of a nation's economy. This relation has been weakening somewhat since 1960 (1960: r-squared

= 0.914, n = 105; 1991: r-squared = 0.734, n = 134), but it remains essentially linear" (Grimes and Timmons 1995, 2). We report a similar finding. Column 1 of Table 3.1 shows a significant association between GDP per capita (logged) and annual releases of toxic emissions.

What is best, then, for lowering pollution rates? The issue is not totally clear, as there are two factors involved in the production of pollutants: the toxic rate per unit of industrial activity and the level of industrial activity. If industries become more and more efficient then the amount of pollutants per unit of industrial activity will continue to decrease. But if these units increase—if growth continues unabated—then while the proportions may be smaller, the overall volume will still increase. The overall amount of toxic pollutants and CO_2 emissions will continue to grow even though the world's high-income countries are becoming more efficient. Limiting unrestrained economic growth has long been a position in environmental circles, but to people of the South, who have yet to acquire the benefits of such unrestrained development, this may be a difficult policy to implement. Efficiency gains seem to also have their limits. While great advances have been made, there is no doubt a limit that will be reached, and if growth continues then the overall volume of pollutants will also increase.

Political Development

Political development (democracy, regime openness, political rights to assemble, etc.) is also associated with a civic culture more favorable to environmental protection and regulation. The democracies of the developed world remain a source of environmental initiatives in regulation, public opinion, and green social movements. In general, pro-environmental attitudes, activities, and organizations provide a potential source of political pressure upon both capital and the state to introduce cleaner technologies and enforce laws limiting toxic pollutants from the production process. The openness of the political system (our democracy variable) should allow more inputs from environmental pressure groups and public opinion and therefore, other things being equal, lower rates of toxic pollution.

To test this hypothesis we enter into a regression equation the Jaggers and Gurr index of democratic openness, GDP per capital as a control for level of development, and data on the annual release of toxic emissions. As column 2 of Table 3.1 shows, there is a statistically significant ($p < .05$) association between democratic openness and toxic emissions, controlling for GDPpc.[2] In effect, the more politically open the country the higher the rate of toxic emission. We know democracy measures and development indicators (GDP) are correlated such that it may be possible that the democracy effect is capturing some of the development effect. (In our analysis the r^2 of GDPpc and democracy was 0.66.)

Dependency

Within the world-system literature there have been a great number of dependent variables that have been shown to be associated with dependency, from

Table 3.1
Regression Estimates of Logged Annual Releases of Toxic Emissions as a Function of Democracy and Dependence

Independent Variables	Dependent Variable: Logged Annual Releases of Toxic Emissions				
	(1)	(2)	(3)	(4)	(5)
GDP per capita (logged)	1.70*	–0.01	0.72	0.53	0.44
	(0.23)	(0.07)	(0.41)	(0.45)	(0.35)
Democracy	...	1.77*	...	...	...
		(0.31)			
Direct Private Foreign	...	...	5.51*	...	6.31*
Investment			(0.00)		(0.00)
% Single Commodity	...	...	...	–0.03*	–0.04*
Export				(0.01)	(0.01)
Constant	–8.60*	–9.0	–2.43	0.88	1.16
	(1.86)	(2.24)	(3.12)	(3.5)	(2.74)
R-Square	0.46	0.48	0.32	0.27	0.57
N	68	66	31	28	28

$^*p <.05$ Note: Standard Errors are in parentheses.

slowed economic growth to changing fertility rates. When toxic pollution is the dependent variable in such an analysis there are a number of hypotheses linking core-periphery relations and dependent development to toxic emissions. The classic product cycle may very well result in a shift of older, more polluting industrial technologies to peripheral/semiperipheral countries. It is also possible that stricter environmental laws in the developed countries force multinational corporations to relocate their more polluting plants as subsidiaries in developing countries which have less strict laws. To examine the possible relationship between dependency and toxic emissions we look at the peripheral and semiperipheral states that comprise our sample.[3] We utilize two indicators of dependence: direct foreign investment and commodity concentration. The results are presented in Table 3.1. Column 3 of Table 3.1 shows a positive significant ($p <.05$) association between direct private foreign investment and toxic emissions, controlling for GDPpc.[4] The more foreign investment, the more toxic emissions. This supports the hypothesis linking dependent development and toxic emissions. Unpacking this statistical association, though, is a more complex issue. There are at least two possible explanations of this finding. It may be that direct foreign investment stimulates more economic activity and that in turn generates more toxic emissions. Evidence against this interpretation is the fact that direct foreign investment is significant even while controlling for level of development. It is not just that more developed countries have more pollution, but holding constant such development, foreign investment is significantly associated with toxic emissions. The second interpretation is that this investment involves plants, technologies, and production processes that emit more toxins. Again, the issue is complex. It may be that more toxin-emitting plants that don't fit under more stringent environmental laws in core countries are moved abroad, or it may be that the type of industry that is built in the periphery/semiperiphery just happens to have a higher rate of toxic emissions.

Column 4 of Table 3.1 shows a significant negative effect of commodity concentration on toxic emission, again controlling for level of development. Column 5 of Table 3.1 places both foreign investment and commodity concentration in the equation at the same time and both remain significant. Foreign investment is still positively associated and commodity concentration negatively associated with toxic emissions. What is interesting here is that while foreign investment and commodity concentration are both standard indicators of dependency, their meanings shift when the object to be explained is environmental. The negative association of commodity concentration means that the more a country's exports are in a single commodity, the less toxic emissions they generate. Again, this is complex. To the extent that commodity concentration reflects an underdeveloped condition of exporting raw materials, then an economy more dependent on raw material exports will have less industry and less toxic emissions. (Obviously certain kinds of raw material exports, like minerals or oil, may involve processing that emits toxins.)

In general, then, the dependency model is supported, although the implications are more complex and point us back to some of the issues raised earlier about different interpretations of environmental justice. The positive effect of foreign investment, controlling for level of development, points toward a deleterious effect of foreign investment for toxic emissions in developing countries. The negative effect of commodity concentration means the more export in a single commodity, the less toxic emission. What these findings highlight are the two definitions of environmental justice we began with. Creating a diversified export portfolio, such that one is not dependent upon any one commodity export, is a path to economic growth, development, and upward mobility in the world-system—all human positives. Yet, this is exactly the path associated with increasing toxic emissions, an environmental (and human) negative. What is good for human development (growth, commodity diversification, development) does not seem good for what could be called environmental development (toxic emissions).

From this point of view introduction of the idea of justice for the environment complicates the moral picture of the development process. We know development has its costs. In recent years the environmental damage seen in Eastern Europe and the Soviet Union have been testaments to the side effects of rapid heavy industrialization. There is also the question of the right of the developing world to manage its environment and natural resources. Environmentalists from developed countries attempting to halt development of the Amazon basin are accused of imperialism by interfering in the rights of local peoples to manage their environmental heritage.

This collision between environmental and human rights does not at present have an obvious solution, although ideas of sustainable development are efforts to provide one. The reduce-consumption, lower needs/wants, less-is-more, small-scale, localism, bioregionalist philosophy seems a solution whose appeal is limited to the developed world. Having much allows one to consider downsizing. Having little makes such reductionism less tenable.

SUMMARY

The data analysis showed mixed results. The democracy indicator was positively associated with emission rates even while controlling for level of development, which does not support the idea that democratic openness allows more pro-environmental opinion to influence government and industry to lower pollution rates. The dependency effects were more clear. Direct private investment is positively associated with, and commodity concentration negatively associated with, toxic emissions. But this raises the two meanings of environmental justice once again. Less development is less emissions. Bad for humans, good for the environment. Resolving the competing interests of humans and the environment remain a major issue.

NOTES

1. Ideas of societal dependence on natural ecosystems and environmental economics are movements in a similar direction. See Daily (1997) and Costanza, Seguro, and Martinez (1996).

2. The differing N's in columns 1–5 reflect missing data on the different variables.

3. The countries used in the regression equations of columns 2–5 in Table 3.1 are marked in the appendix with an asterisk.

4. In a separate analysis we logged direct private investment. The results were not significantly different.

REFERENCES

Bergesen, Albert J.. 1995a. Deep Ecology and Moral Community. In *Rethinking Materialism: Perspectives on the Spiritual Dimension of Economic Behavior,* ed. Robert Wuthnow, 193–213. Grand Rapids, MI: Erdmanns.

———. 1995b. Eco-Alienation. *Humboldt Journal of Social Relations* 21: 1–14.

Bunker, Stephen G. 1984. Modes of Extraction, Unequal Exchange, and the Progressive Underdevelopment of an Extreme Periphery: The Brazilian Amazon. *American Journal of Sociology* 89(5): 1017–64.

———. 1985. *Underdeveloping the Amazon: Extraction, Unequal Exchange, and the Failure of the Modern State.* Urbana: University of Illinois Press.

———. 1994. Flimsy Joint Ventures in Fragile Environments. In *States, Firms, and Raw Materials: The World Economy and Ecology of Aluminum,* ed. B. Barham, S. G. Bunker, and D. O'Hearn, 261–96. Madison: University of Wisconsin Press.

Bunker, Stephen, and Paul S. Ciccantell. 1995. Restructuring Space, Time and Competitive Advantage in the Capitalist World-Economy: Japan and Raw Materials Transport after World War II. In *A New World Order? Global Transformation in the Late Twentieth Century,* ed. David A. Smith and Jozsef Borocz, 109–129. Westport, CT: Greenwood.

Burke, L. M. 1993. Race and Environmental Equity: A Geographic Analysis in Los Angeles. *Geo Info Systems* 3(9): 44–50.

Burns, Thomas J., Edward L. Kick, David A. Murray, and Dixie A. Murray. 1994. Demography, Development and Deforestation in a World-System Perspective. *International Journal of Comparative Sociology* 35: 221–39.

Chew, Sing C. 1996. Wood, Environmental Imperatives and Developmental Strategies: Challenges for Southeast Asia. In *Asia—Who Pays for Growth? Women, Environment, and Popular Movements,* ed. Jayant Lele and Wisdom Tettey, 206–26. Brookfield, VT: Dartmouth.

Costanza, R., O. Segura, and J. Martinez, eds. 1996. *Getting Down to Earth: Practical Applications of Ecological Economics.* Washington, DC: Island Press.

Daily, Gretchen C., ed. 1997. *Nature's Services: Societal Dependence on Natural Ecosystems.* Washington, DC: Island Press.

Dietz, Thomas, and Eugene A. Rosa. 1997. Effects of Population and Affluence on CO_2 Emissions. *Proceedings of the National Academy of Sciences of the USA* 94 (Jan.): 175–79.

Flavin, Christopher, and Alan Durning. 1988. Raising Energy Efficiency. In *State of the World 1988,* ed. Lester R. Brown et al. New York: Norton.

Glickman, T. S., and R. Hersh. 1995. Evaluating Environmental Equity: The Impacts of Industrial Hazards on Selected Social Groups in Allegheny County, Pennsylvania. Discussion Paper 95-13. Washington, DC: Resources for the Future.

Grimes, Peter, and J. Timmons. 1995. "Carbon Dioxide Emissions Efficiency and Economic Development." Department of Sociology, Johns Hopkins University.

Hettige, Hemamala, Robert E. B. Lucas, and David Wheeler. 1992. The Toxic Intensity of Industrial Production: Global Patterns, Trends, and Trade Policy. *American Economics Review* 82(2): 478–81.

IMF. 1991, 1995. *Balance of Payments Statistics Yearbook.* Washington, DC: IMF.

Jaggers, Keith, and Ted Robert Gurr. 1996. *Polity III: Regime Type and Political Authority, 1800–1994.* Ann Arbor, MI: Inter-university Consortium for Political and Social Research.

Kick, Edward L., Thomas J. Burns, Byron Davis, David A. Murray, and Dixie A. Murray. 1996. Impacts of Domestic Population Dyamics and Foreign Wood Trade on Deforestation: A World-System Perspective. *Journal of Developing Societies* 12: 68–87.

Low, Patrick, and Alexander Yeats. 1992. Do "Dirty" Industries Migrate? In *International Trade and the Environment,* ed. Patrick Low, 89–104. World Bank Discussion Papers no. 159. Washington, DC: International Bank for Reconstruction and Development/World Bank.

Lucas, Robert E. B., David Wheeler, and Hemamala Hettige. 1992. Economic Development, Environmental Regulation and the International Migration of Toxic Industrial Pollution: 1960–88. In *International Trade and the Environment,* ed. Patrick Low, 67-86. World Bank Discussion Papers no. 159. Washington, DC: International Bank for Reconstruction and Development/World Bank.

Pollock, P. H., and E. M. Vittas. 1995. Who Bears the Burdens of Environmental Pollution? Race, Ethnicity, and Environmental Equity in Florida. *Social Science Quarterly* 76(2): 294–310.

Roberts, J. Timmons. 1996. Predicting Participation in Environmental Treaties: A World-System Analysis. *Sociological Inquiry* 66: 38–57.

Roberts, J. Timmons, and Peter E. Grimes. 1997. Carbon Intensity and Economic Development 1962–91: A Brief Exploration of the Environmental Kuznets Curve. *World Development* 25(2): 191–98.

Rudel, Thomas K. 1989. Population, Development, and Tropical Deforestation: A Cross-National Study. *Rural Sociology* 54: 327–38.

Smith, David A. 1994. Uneven Development and the Environment: Toward a World-System Perspective. *Humboldt Journal of Social Relations* 20: 151–75.

Summers, Robert, and Alan Heston. 1991. The Penn World Tables (Mark 5): An Expanded Set of International Comparisons, 1950–1988. *Quarterly Journal of Economics* 106: 327–68.

Szasz, Andrew, and Michael Meuser. 1997. "Environmental Inequalities: Literature Review and Proposals for New Directions in Research and Theory." *Current Sociology* 45(3): 99-120.

Szasz, Andrew, M. Meuser, H. Aronson, and H. Fukurai. 1993. "Demographics of Proximity to Toxic Pollution: The Case of Los Angeles County." Department of Sociology, University of California, Santa Cruz.

UNCTAD. 1983, 1984, 1986, 1988, 1990. *Yearbook of International Commodity Statistics.* New York: United Nations.

APPENDIX

Country Sample for Regression Analysis in Table 3.1.

Australia
Austria
Bangladesh*
Botswana*
Cameroon*
Canada
Central African Republic*
China*
Colombia*
Congo*
Costa Rica*
Cote d'Ivoire*
Cyprus*
Denmark
Dominican Republic*
Ecuador*
El Salvador*
Ethiopia*
Fiji*
Finland
France
Germany
Ghana*
Greece
Guatemala*
Hong Kong
Hungary
India*
Indonesia*
Ireland
Italy
Japan
Kenya*
Korea, Republic of
Kuwait*
Libya*
Luxembourg
Madagascar*
Malawi*
Malaysia*
Malta*
Mauritius*
Morocco*
Netherlands
New Zealand
Nigeria*
Norway
Pakistan*
Panama*
Papua New Guinea*
Philippines*
Poland
Portugal
Rwanda*
Senegal*
Singapore
South Africa
Spain
Sri Lanka*
Sweden
Syrian Arab Republic*
Thailand*
Tunisia*
Turkey*
United Kingdom
United States
Uruguay*
Venezuela*
Yemen*
Zimbabwe*

* = countries employed in regression analysis of columns 2–5 of Table 3.1

4

Extending the World-System to the Whole System: Toward a Political Economy of the Biosphere

J. Timmons Roberts & Peter E. Grimes

INTRODUCTION

Until only very recently, writing in the world-system tradition focused so closely on the global social structure of accumulation (that is, crudely, how profits are made and by whom at whose expense) that it overlooked the natural environment upon which it depends. Like much of social science, world-system theory has implicitly taken what Dunlap and Catton (1994) called the "human exemptionalist" approach—that humans are exempt from ecological laws affecting other species. In spite of nearly two decades of neglect of the physical environment, we believe that world-system concepts hold great and virtually untapped promise for addressing global environmental issues.

We here explore how the core insights of world-system theory might inform environmental understanding. Our approach is to first look at why world-system theory hasn't addressed environmental issues, then in the middle part of the chapter we explore the value of world-system theory's central tenants and main themes in addressing global environmental issues. We examine in sequence the environmental utility of four world-system tenants and six questions which have received the greatest attention from its practitioners. The core tenants we examine are world-system theory's explicit and defining globalism, its materialist perspective, its historicism, and its structuralism. We explore the environmental implications of world-system theory's attention to secular trends and cycles in global capitalism, its attention to identifying key economic and political actors, to the causes of war, to the exploitation of peripheral (poor) nations, and finally of its concern with socialism and the transition of the former Soviet bloc nations. There are others we could have added to the list. World-system theory is a huge field, and it is difficult to summarize its potential contribution to environmental understanding, but we attempt to do so without stereotyping. In a concluding section, we assess the weaknesses and limitations of the world-system approach and propose a discussion of possibilities for future world-system analysis of pollution emissions and resource depletion.

At least as narrowly defined, we do not think world-system theory has all the answers, but it does have great potential. When integrated into a wider political economy that addresses a major shortcoming by considering culture seriously, perhaps world-system theory can play a central role in integrating environmental sociology.

WHY WORLD-SYSTEMS THEORY MISSED THE GREEN WAVE

For a variety of reasons, the *zeitgeist* and political agendas of the times that gave birth to world-system theory did not place a high priority on environmental concerns. However, its quick acceptance (in the United States) following the publication of Wallerstein's *Modern World-System I* in 1974 revealed the fertile ground created by the prior decade of radicalization among intellectuals (e.g., the Civil Rights movement, opposition to the Vietnam war). The popularity of world-system theory also grew out of a questioning of the "Development Project," as formulated in Washington after World War II, in which Third World nations were placed in roles of recipients of U.S. aid and purchasers of U.S. products (McMichael 1996).

A profound split among Left academics became apparent, even in those early years of world-system theory. While the political struggles of the 1960s and early 1970s were generating a "New Left" and thereby laying the groundwork for the eager acceptance of world-system theory, some disenchanted intellectuals sought escape by going "back to the land." That is, there was a perceived choice between fighting for social justice or for environmental protection, and an unfortunate cultural split developed between anti-imperialism/Vietnam war politics and those drawn toward a more voluntaristic, lifestyle change approach (e.g., "back to the land"). Both movements are attempting to bridge this gap even today, and other nations often face the same problem in one form or another.

How, then, did world-system theory miss the "green wave" of environmental sociology? Why did its original formulations ignore environmental concerns? Perhaps Buttel (1978) and Dunlap and Catton (1994) are correct in pointing out how environmental sociology failed in its early hopes to integrate environmental concerns into the broader discipline. On the academic front, world-system theory reached its apex of popularity in the United States around 1983, when researchers of several stripes tried to tie their real interests to the fad by including the phrase "World-System" somewhere in their paper titles. But between 1985 and 1990 world-system theory was beginning to lose favor, retreating from the multiple attacks of volunteerists, historicists, feminists, and postmodernists, who collectively shifted the attention of social scientists away from the macro-level and meta-narratives such as theories of development and toward the micro-level politics of identity. By an accident of timing, these years between 1983 and 1993 were precisely the years when popular knowledge of the ozone hole, rainforest destruction, and global warming were growing. Had it not been for the onslaught of attack and intellectual intimidation dampening the receptivity to world-system analysis at that very time, we suspect that world-system theory would have turned to the environment a decade ago.

As it was, in the mid-1980s Stephen Bunker, working on extractive economies such as the Brazilian Amazon, was a lone voice in the proverbial wilderness—attempting to combine environment and energy issues with the central issues of world-system theory (1984, 1985). His book *Underdeveloping the Amazon* faulted both Marxian and Neoclassical economics for placing the origin of all value in human work and in production, thereby failing to understand the role of naturally occurring resources in creating value or "rent." He proposed a reexamination of the concepts of unequal exchange to include the extraction of energy and material resources from the peripheries of the world-system and described the social and environmental consequences of the typical boom-bust resource cycle. Freudenberg (1981) and Gramling and Brabant (1990) have long examined social structures of resource boomtowns, but there was little crossover of these insights into world-system circles. Rather, J. T. Roberts (1992, 1995a; Roberts and Dodoo 1995), Barbosa (1993, 1996), and Ciccantell's (1994) work have most closely followed Bunker's lead.

There were some important efforts made, but little crossover in the other direction from environmental sociology to the core of the Political Economy of the World-System (PEWS) group. Instead, environmental sociology took as needed from neo-Marxist and world-system analysis and developed on its own, while world-system theorists continued to ignore environmental thinking. For example, Schnaiberg's important 1980 contribution, *The Environment,* while incorporating political economy explicitly and describing the "treadmill of production" which pushes governments and firms to ever increase their environmental impact, was ignored in the world-system field. Rudel's work (1989) combining detailed analysis of deforestation in Ecuador with political economy perspectives, such as Logan and Molotch's concept (1987) of the "growth machine," gained little attention within PEWS. Perelman (1977), Burbach and Flynn (1980), Buttel (1987), and Kloppenburg's (1988) work on the international political economy of agricultural systems had explicit implications for biodiversity and other environmental issues. Likewise the cross-national political economy of energy consumption work by Buttel (1978) and Mazur and Rosa (1974) was never widely used among world-system researchers. However none of the environmental issues became central to the 1980s work of world-system theorists, even those that were focused on agricultural systems, until the mid-1990s (see above and McMichael 1996). The same is true of O'Connor's essays and efforts in creating the journal *Capitalism, Nature, Socialism.* Rather, many of the questions of how the world-economy and the environment were related were being taken up by human geographers and some economists (e.g., Dicken 1992; Ayres 1989; Simonis 1989). Some adopted world-system theory concepts while adapting methodologies to their disciplinary styles (Chew 1995, 1997).

The last few years, however, have seen more attention to the environment within world-system theory. Some examples of the new "better, greener world-system theory" boom can be described briefly, and we will return to many of them later. Chew (1995) has recently attributed the decline of large empires throughout history to massive deforestation and land degradation. Sanderson (1995) also explicitly incorporates environmental factors to explain the succession of social forms throughout history, as do Chase-Dunn and Hall (1997) in their most recent book. Frey (1993a, 1993b) has presented some work on the political economy of the hazardous waste trade building on an earlier influential piece with Covello (1990). Barnham, Bunker, and O'Hearn (1994) and Ciccantell (1994) continue to work on the environmentally important sector of mining and metals, but often they have not developed the explicitly environmental links (Bunker did so in 1985, 1989). Gellert (1996) is working on the lumber industry in East Asia, but like Bunker's work on extractive commodities, Gellert's has apparently not clarified the larger environmental implications. To the contrary, Barbosa (1993, 1996) is publishing on how the struggle over the Amazon fits into global ecopolitics. Yearley's recent book (1996) provides a sharp critique to the literatures on both globalization and global environmental issues.

Finally, our own work spans three areas. Roberts (1994, 1996a) has summarized the Brazilian scene and drawn a wide portrait of Latin American restructuring and its environmental implications. He has also tested the ability of world-system theories and methodologies to predict which nations will participate in international environmental treaties (1996b). Together, Grimes and Roberts's work has used world-system theory and methods to explain national variations in greenhouse gas emissions, both presently and historically (1995, 1996). We will return to these authors throughout the next section, but because there is quite little by way of basic theoretical foundations bridging world-system theory and the environment, we should make clear that what follows is more an exploration than a review of previous works.

HOW THE CORE TENETS OF THE WORLD-SYSTEMS PERSPECTIVE COULD INFORM ENVIRONMENTAL SOCIOLOGY

In this section, we consider each of what we consider four of the central world-system theory tenets and the insights each could provide to our understanding of humanity's relations to its natural environment. After these, we examine the six questions which have received the most attention from world-system researchers.

The Core World-System Theory Idea

First, we consider the core world-system theory idea: that we must look at everything. Braudel (1981) once wrote to the effect that no description can even begin to lead to a valid explanation if it does not effectively encompass the whole world. Some of us might find this requirement for any explanation impossible, but it remains an ideal for world-system researchers. This proposition implies that we cannot look at one community's or nation's relation to the environment, but that we must also understand how they are linked to larger social organizations and the global society, and how they collectively influence and are influenced by the environment. Two examples may help illuminate the importance of this simple point. First, the ecologically damaging effects of producing monocultural crops for export have been repeatedly well documented (e.g., Perelman 1977; Wright 1990; Barkin 1995). However to understand why poor and rich countries alike decide to devastate their soils and contaminate their rivers with agrotoxics, we need to understand the pressures they are under to compete in a global economy. Second, localities in the world with relatively greater environmental awareness and regulations on industry often feel pressures to lower their requirements in order to keep their tax base and jobs and to compete for new plant sitings (Kazis and Grossman 1991; Gould 1994; Reed 1992). Further, with free trade agreements such as the GATT/WTO, it can be considered protectionism for one nation to have unusually high environmental requirements of products produced or consumed there (Roberts 1996c). By integrating production from widely dispersed locations into one coherent global system, the capitalist world-economy likewise scatters the ecological costs of that system unequally throughout the globe.

It is easy to say that the need to raise export earnings to pay off the national debt is driving reckless resource use around the world (e.g., Covello and Frey 1990; Reed 1992; Gould 1994). However, understanding how the global capitalist system is changing, and the direction of likely future changes, requires a sophisticated understanding of its past and emergent mechanisms. We return to secular trends and cycles in the system later, but for the present the point is that good environmental policy requires an understanding of the whole global system and how it can support or subvert local efforts at environmental protection. Elsewhere, Grimes and Roberts (1995) and Roberts (1996a; 1996c) have argued that the globalizing economy holds both perils and prospects for environmental protection. While GATT/WTO-type treaties are potentially driving down the environmental sovereignty of nations, the emergence of global environmental standards holds important potential to keep firms from fleeing to so-called "pollution havens" (Roberts 1996c).

Ironically, while they built largely on Braudel's shoulders, world-system theorists who came later failed to look at nature. As Chew (1997) recently pointed out, the world-system approach at first included natural causative forces. Braudel, especially in his earlier works, such as *The Mediterranean* (1972), spent much time detailing how the specifics of that society and economy grew from the nature of that land. He continued that focus in his influential three-volume work, *Civilization and Capitalism* (1981), especially in Volume 1 (*The Structures of Everyday Life*). Chew takes Wallerstein (until 1996), Amin, Frank (until his 1991 and 1993 work with Gills), Chase-Dunn (until the 1996 book with Hall), and Wolf to task for treating nature as external and as a backdrop to the true engine of change: social relations in general and capitalism in particular. Rather, Chew points out that "the social and ecological (natural) worlds interact in a dialectical fashion whereby Nature's rhythms also impact on the dynamics of social-economic life. For example, changes in climatological trends . . . will impact on crop harvests, which in turn will determine grain prices or the migration of people" (1997, 13). Meanwhile human-driven "degradative effects on Nature . . . in turn loop back to impact on the dynamics of social-economic life of the world system." While this point may seem vague and obvious to anyone who has studied human-environment interactions, it represents an important step in this case. World-system theory, being built on Marxian analysis, saw the ultimate demise of the global capitalist system coming from the contradictions in the capitalist model itself (especially accumulation crises, working class struggles, and the inevitable socialist revolution; Chew 1997, 22). However world-system theorists are finally beginning to see that the critical contradiction is more likely to be the final overtaxing of the global ecological base which has supported capitalism's rise.[1]

A Materialist Theory of Capitalism and Development

Second, world-system theory is a materialist theory of capitalism and development. Its materialism is a source of strength for environmental analysis, and simultaneously an important shortcoming, which we discuss further below. World-system theory is, of course, a subfield of political economy: it is a theory of capitalism and development. World-system theory attempts to explain how societies change, building on neo-Marxist theories of imperialism, dependency, and theories about how labor is organized in different societies (Grimes 1988; Chilcote 1984; Brewer 1980). The central conception here is that human actions based on subsistence (as socially defined and ever changing) and profit making are the core of the economy and that political and cultural structures are built around the needs of the economy (Marx 1967). World-system theory emerged in opposition to cultural and stage theories of development, especially modernization theory. This explains perhaps why world-system theorists have "foregrounded" economic structures as determinant while expressing an aversion to cultural explanations. The world-system perspective has been roundly critiqued for attempting to "read the state off the economy" (Skocpol 1985; Evans 1979; Evans, Rueschemeyer, and Skocpol 1985) and for ignoring cultural factors such as gender and culture, but the main protagonists have proven willing to incorporate many of these critiques in later work (for example, PEWS has held conferences and published volumes on the state and how households fit in the world economy).[2]

Returning to world-system theory's deep materialism, the environmental implications are clear: production and consumption directly affect the biosphere and cannot be understood without understanding the structure of the world economy. If many or even most human activities are economically motivated, or if the most damaging human actions for the environment are those based on human subsistence and economic decision making, then a materialist core for environmental theory is appropriate (see, e.g., Schnaiberg 1980; Foster 1994; O'Connor 1989; Harper 1996). Again, the uniquely world-system contribution to that discussion is that much production activity is either directly for trade or designed to facilitate trade and or make it more profitable (e.g., military foreign policy). Production and consumption do indeed seem heavily guided by the world economy, by the needs of markets which are increasingly distant from the site of production.

A Long Historical View

Third, world-system theory takes a long historical view to understand the development of our current social/economic systems. As Shannon (1996) points out in his useful book *An Introduction to the World-System Perspective,* world-system theory is more historical than most of U.S. sociology and perhaps most of U.S. social science. Its most visible theoreticians—Wallerstein, Braudel, Chase-Dunn, Sanderson, Frank—all have taken deeply historical approaches in much of their writing. Wallerstein and Braudel both began their accounts with medieval Europe. Mann, Frank, Chase-Dunn, and Sanderson went even further to describe social evolution over millennia, the latter two building their explanations of the transitions on a modified form of the Lenskis' technology-based account (Mann 1986; Frank 1978; Chase-Dunn 1989; Chase-Dunn and Hall 1996; Sanderson 1988, 1995). Incorporating archaeological information, these authors have even explored the evolution of "pristine" states, long-distance trade, and early subsistence systems of different world-systems in the period before written history (Chase-Dunn and Hall, 1991).

One central goal in these accounts has been to explain the key social transitions: from hunter-gatherer band to horticultural chiefdoms, agricultural tributary states/empires, and, finally, contemporary international industrial capitalism. Much debate has centered over when capitalism emerged and its relation to noncapitalist societies. A lingering disagreement exists over whether capitalism should be defined by the world trading system and market (Wallerstein 1974; Frank 1969) or by the relations of labor such as slavery or wage work (production; Laclau 1971; Brenner 1977). Competing definitions of capitalism have allowed a thriving debate over when global capitalism emerged (see the summary in Chase-Dunn and Grimes, 1995). Frank and Gills (1993) now argue that capitalism has been around for five thousand years, not the five hundred that Wallerstein claimed (see recent pieces in the *Humboldt Journal and Review* and the ongoing debate on the electronic world-systems network listserver WSN@csf.colorado.edu). This seemingly arcane point affects how one bounds world-systems across both time and space. However, it does not change the fact that all major empires and civilizations, no matter how one labels their mode of accumulation, were brought down at least partly by land degradation. For example, many Greens and environmental sociologists blame capitalism for the current ecological crisis. However as Harper (1996) and several others have pointed out, environmental crises have surfaced in societies of many types throughout history. The policy implication is massive: is capitalism inevitably caught on a treadmill which drives it ever forward (Schnaiberg 1980), or can it be made sustainable? Among sociologists, world-system theorists are well positioned to answer questions about whether there have been truly stable capitalist societies and how they were successful in the face of expansionist states and markets.

Second, world-system theory's focus on the relation between the capitalism of early Europe and other societies and their modes of production highlighted the pivotal importance of the colonial relations which have conditioned most nations' social and environmental situations since then. That is, we cannot understand a nation's current position without understanding its colonial past. Schafer's (1994) pioneering work on *commodities* and development points out the limitations placed on nations who rely heavily on one type of product for the majority of their exports. He relates a nation's export commodities to its available paths of development, demonstrating that the historical choices of production technologies made by former colonial powers continue to impose *political* constraints on production technology today. Bunker and colleagues have taken up this approach in an extended study of the aluminum industry and markets around the world, and its effects on local producing nations (Barnham, Bunker, and O'Hearn 1994). The environmental implication of this "dead hand of the past" is that nations heavily dependent on the export of commodities such as agricultural products, oil, or even low-priced manufactures often have weak civil societies and states dominated by the exporting sector's elite, largely dependent upon them for state revenue. This often corresponds with weak or nonexistent environmental movements and relatively little state autonomy or strong constraints on the state, as it avoids making the "tough decisions" to break the resource exploitation habit and move toward higher value-added products (Bunker 1985, 1994; Ranis 1990; Roberts 1996a). The cycle of resource dependency and environmental degradation is likely to continue until all the resources have been exploited, when crisis and collapse leaves an even weaker state and little prospect for positive civil reform.

Structural Positions in a Global Stratification System

Third, world-system theory holds that nations occupy structural positions in a global stratification system. Like class structures within nations, a few nations move up and down, but the structure remains intact over time. Also like most national stratification systems, the majority of nations remain trapped at their current level in the global system. The relevance of attempting to characterize this global system of inequality becomes clearer if we remember Indhira Gandhi's point at the 1970 Stockholm conference that the pollution of poverty differs from the pollution of wealth. For example much of our own previous work has shown that the relative wealth and world-system position of nations is an excellent (but not at all complete) initial predictor of the level or type of pollution it is likely to be creating (Grimes, Roberts, and Manale 1993; Roberts and Grimes 1997), as well as its level of commitment to international environmental agreements (Roberts 1996b).

It was mentioned above that many remnants of modernization theory persist in Washington policy circles today. One derivative argument which has gained favor in the 1990s is the "Environmental Kuznets' Curve." When one plots some types of pollutants such as levels of urban smog against a nation's GNP per capita, an upside-down U-curve is evident. This fact led influential economists Grossman and Krueger (e.g., 1995) and the World Bank in its influential *World Development Report* of 1992 to argue that countries will first get worse as they develop and, after reaching some turning point, will improve their environmental performance. The argument is based on Maslow's "hierarchy of needs" and the "post-materialist" hypothesis (see Inglehart 1995; Roberts and Grimes 1997). On the face of it, the argument makes sense: that nations will ignore pollution controls until they have dealt with basic human needs and only then will they begin to care about quality of life issues like clean air and water. As then-president of Mexico Carlos Salinas is reported to have said on the U.S. PBS show "The McLaughlan Group," "We have to pollute now, to develop and deal with pollution later." Based on the pattern for several pollutants, Grossman and Krueger put the turning point at about U.S. $8,000 per capita. The policy implication is that nations need economic growth *first* to reach environmental protection later.

However, based on the repeated findings of the world-system theory school, we argued in a recent *World Development* article (Roberts and Grimes 1997) that assuming that things will get better for the world's poor nations is extremely perilous and in fact historically counterfactual. Research presented by Sanderson (1995) demonstrates that, contrary to the assumptions of the modernization theory underlying the Kuznets' curve, the gap between countries has grown geometrically over the past century. Some environmental effects have Kuznets' curve-type relationships with national wealth, while others increase linearly or even more quickly as countries get richer. In our examination of the historical trend over 30 years for national carbon intensity, it becomes clear that the environmental Kuznets' curve does not represent an historical trend, but is merely a cross sectional pattern which emerged in the 1980s and which is actually likely to worsen (Roberts and Grimes 1997). These findings are consistent with the expectations of world-system theory that nations are generally trapped in the global stratification system. Certainly there are cases of mobility, but mobility has been in both directions (up and down), and in general poverty persists. Even the World Bank in its 1995 report admitted that (in spite of 50 years of the "Development Project") inequality between nations has in fact worsened sharply. The clear implication is that environmental problems cannot be left to resolve themselves by economic growth alone: without massive changes in the system there will always be poor nations being exploited who will therefore have to exploit their natural environment. Environmental protection must be addressed at all levels of development.

Aren't the movement of factories to poor countries, "globalization," and the "New International Division of Labor" bringing development that will help those nations move up in wealth and relative position in the global hierarchy? Again, empirical world-system research suggests not. Even as factories—once associated with development and riches—move to the semiperiphery and periphery of the world system, they are no longer necessarily bringing wealth (Dicken 1992; McMichael 1996). The reasons lie in an area well explored by a subfield of world-system theory: the commodity chain approach. Gereffi, Miguel and Patricio Korzeniewicz, and several other contributors to their volume have pointed out (see Gereffi and Korzeniewicz 1994) that nations in the core continue to control production and marketing of most products and so capture most of the profits from their sale. The commodity chain approach moves us along the road towards understanding which nations are more likely to move up economically into higher standards of living. Combining some insights, those nations who soon might be able to spend more on environmental protection are those which have been selling valuable commodities on the world market and which have captured some higher-value portions of the commodity chain (Shafer 1994; Gereffi and Korzeniewicz 1994). Put another way, as the profits from a given method of production decline with age, such methods are exported out of the core (Dicken 1992). So industrialization outside of the core does not mean advancement, but only that industrial methods are no longer the sites of high profit that they once were a generation ago.

How strictly do types and amounts of pollution vary by a nation's position in the world-system? Insofar as wealth is correlated with pollution, the world-system perspective has important contributions to make. Because it is a key to empirical research utilizing the world-system approach, we discuss some details of the debate over the best way to categorize nations in the world stratification system. It is helpful to remember that while the debate seems sometimes a case of minutia, the key question is which nations are in a powerful enough position to have substantial control over the sources of pollution in their territories and even around the world.

Some authors have argued that periphery, semiperiphery, and core are qualitatively different roles in the world political and economic systems (following Wallerstein 1974, 1979). Further, each nation generally is placed in just one of these roles. This leads to some debate about marginal cases: everyone agrees that the United States, Germany, and Japan are core nations, but what about wealthy but weak nations like Sweden, Spain, or Luxembourg, or powerful, large but poor countries like China, Brazil, and India? Is political or economic power the most important factor? Another approach has been to conceptualize the global stratification system as a continuum and to rank nations in their total power or position in world trading networks, for example. Many researchers simply use gross national product per capita, an approach they ironically share with many modernizationists (e.g., Arrighi and Drangel 1986). This approach has the distinct advantage that data is available for many nations over many years. This is not the case with more subtle classification systems. Smith and White (1992) took a more elaborate approach, classifying nations in 1970, 1980, and 1985 by network cluster analysis of the relative value of the products they imported and exported. Dutch researcher Terlouw (1992) compared several approaches to classifying nations. He first completed a factor analysis of how nations were classified in nine major world-system theory works. These rankings were placed on a scale and compared with an index of a nation's overall presence in world trading networks—their exports as a percent of global exports. Another compromise was produced by Grimes, who combined these two indexes into one hybrid World-System Position (WSP) index (Grimes, Roberts, and Manale 1993). Grimes (1996) compiled a new WSP index for every country for which adequate data was available over the past 200 years (in five-year intervals). First, he calculated for each nation their GDP as a percent of the global GDP. Second, he weighted this GDP/Global GDP measure by that nation's dependence on the most important trading partners and by the relative position of those trading partners in the world-system.

How do these different classification schemes do in separating the "pollution of poverty" from that of wealth? We believe that exploring this question would be a worthwhile empirical exercise with a range of environmental outcomes. Roberts (1996a,b,c) examined the abilities of three indicators of WSP—GDP/cap, Terlouw's consensus index, Terlouw's "presence in global trade" WSP index, and Grimes' hybrid of the two Terlouw positions—to explain whether nations had signed the major environmental treaties promulgated during the period 1963–1987. The WSP indicators were superior to GDP/cap and explained over half of the variance. This supports the idea that WSP and national wealth explain a substantial amount of nations' environmental performance. However in all cases at all levels of income and WSP, there is tremendous variation, which suggests that there is NO necessary relation between a nation's wealth and its pollution and environmental degradation. At every WSP, there is a large scatter of pollution levels, demonstrating that many factors outside of either WSP or GDP/Capita shape the overall pollution profile. That this relation is not cast in stone is shown also by the historical de-linking of economic growth and environmental impacts in some nations (Simonis 1989).

Finally, it is important to note that the world-system approach has led its practitioners to pay special attention to nations in what is called the "rising semiperiphery." These are nations which are making industrialization drives in attempts to gain core status. These also, of course, correspond to areas of looming environmental disasters:

- — Brazil's incorporation of its Amazon frontier to spur its drive to development (Bunker 1985; Barbosa 1993; Roberts 1995b);
- — China's rapid industrialization and its booming need for electrical power, most likely using coal and drastically increasing carbon outputs (e.g., Howard 1993; Smith 1994);
- — Malaysia and Indonesia's logging booms which are driving rapid deforestation there;
- — Eastern Europe and Russia's reckless push for heavy industrialization (Manser 1993).

These are the emerging environmental disasters of our day, and world-system theory can shed some much-needed light on which nations are likely to succeed in those drives to ascend in WSP and which will fail, when, and under what conditions. What is largely lacking are any theories of which types and levels of environmental effects tend to go with WSP ascent and decline.

Having discussed the environmental implications of four of world-system theory's central tenets, we now go on to discuss more briefly six areas which the field has discussed in depth. Again the list is necessarily incomplete.

THE ENVIRONMENT AND WORLD-SYSTEM THEORY'S CENTRAL QUESTIONS

Without being Panglossian, we believe that the attention world-system theorists have paid to several core issues over the years—some of which now appear quite esoteric—can provide crucial insights to long-running debates in environmental studies. We here lay out six core issues/questions of world-system theory and discuss their environmental implications.

Key Secular Trends

Research in the world-system tradition has identified and examined key secular trends in global capitalism. They are surveyed in more thorough detail in Chase-Dunn and Grimes (1995), so will only be touched upon here. These are widespread trends in society such as increasing commodification, the tendency toward the greater proletarianization of the labor force (or at least their incorporation into the world-economy), the expansion of the state, the growth of corporate power and its ability to escape the control of states and civil society, and globalization more generally.[3]

The environmental by-products of many of these trends is enormous. Briefly, *commodification* combined with the technologies of long-distance trade and refrigeration has brought about both a progressive removal of producers from consumers and the increasing specialization of production (monocrop agriculture). Most workers are far removed from the land that sustains them, and virtually everything is produced for profit rather than for use-value. Further, the trend toward what Wallerstein (1989) calls the "commodification of everything" has also included the commodification of nature, accelerating its destruction in pursuit of increasingly rare items (e.g., black rhino horn, ivory, tropical birds).

Following Marx, *proletarianization*—the reduction of what were formerly "professions" with a high degree of job autonomy to de-skilled low-autonomy routines requiring lower education and more amenable to ultimate automation—has been said to drive alienation, inequality, and fatalism (Braverman 1974). More to the point here, it enables automation, which is the substitution of inanimate energy for human labor and which drives up energy consumption and increases toxic effluents.[4] Both of these remove from workers the ability to control the effects of their labor on the environment or on their own health and safety.

Expanding states require expanding militaries and bureaucracies, which in turn are extremely resource- and pollution-intensive (Grimes, Roberts, and Manale 1993). On the other hand, many authors fear that *growing corporate power* and its ability to flee from controlling states and civil groups such as labor and environmentalists may close the last of any real accountability on its part. One example of the effects of *globalization*: rises in the stock markets in Hong Kong drive up the Dow Industrials that in turn spurs interest rate fears and worries about inflation "rearing its ugly head." The U.S. Federal Reserve tightens up its lending and this makes it more difficult for Third World states to gain credit to service their foreign debts. This debt crunch influences decisions about lumber leases and types of mining joint ventures they choose to enter into (Reed 1992). As times get tighter, levels of commitment to environmental protection drop (Roberts 1996b). The ongoing Mexican economic crisis is a prime example of how this system works and how debt and currency issues and the price of a key commodity (oil) can cause deep shocks in an economy and drive down environmental spending (Barkin 1995). A hostile takeover of one company by another using leveraged buy-out stock offerings can compel the purchased target corporation to exploit it workers and resources much more harshly in order to raise the revenue to meet the new debt obligations taken on by the original author of the hostile takeover. When the affected corporate resources include extractive industries or agriculture, the environmental consequences are necessarily accelerated. Hence the global interconnectedness of the world-economy allows for the actions of traders in New York or Hong Kong to have unanticipated collateral effects on environmental degradation everywhere.

Causes and Effects of Cycles of Crisis

World-system theory identifies and examines the causes and effects of cycles of crisis and restructuring in global capitalism. The central idea is that capitalism has some internal contradictions which cause it to periodically stall, reorganize itself, and move forward once again. While many cycles have been found in historical data on prices, production, and worker militance, some are more widely accepted than others. The existence of cycles of various lengths in crucial environmental factors such as temperature and rainfall suggests that cycles in capitalism deserve our attention.

The most important cycles identified by world-system theorists include Kondratieff's 50- to 60-year "long waves," Kuznets' cycles of 15 to 20 years in duration, and Juglar or "regular business cycles" 5 to 8 years in length (reviewed in Chase-Dunn and Grimes 1995). These cycles are not mutually exclusive; all continue to operate simultaneously in a "superimposed" fashion. Little work has been done relating these cycles with different types and intensities of environmental damage (see Chew 1997). However, it merits examination whether there are types of environmental damage we should expect in different parts of the different cycles. This has been seen to be the case with several other social outcomes such as wars and strikes (Goldstein 1988; Chase-Dunn 1989; Chase-Dunn and Grimes 1995). One obvious prediction is that during cycles of economic expansion which correspond with some types of geographical expansion or colonization, certain environmental effects such as land use and energy consumption would worsen considerably. This was true in the colonization of North America (Tucker and Richards 1983) and the more contemporary expansion of the Brazilian state into the Amazon frontier since the early 1960s (Bunker 1985; Wood and Carvalho 1988; Santos 1980). During these expansionist phases, new areas are opened up to exploitation, but many of the worst environmental effects in quantitative terms may await the more intensifying phase which follows.

During down or contractionist phases, exploitation presumably also contracts. Deforestation rates in the Brazilian Amazon, for example, slowed considerably in the 1987–1990 period, partly because of the economic crisis and wild inflation there (Barbosa 1996). However, even with continuing crisis, the rate has now been seen to be back on the rise since 1991, in keeping with the global recovery that started then. Furthermore, with highly mechanized tools of environmental destruction, even while downsizing its workforce, a lumbering or mining company can still cut or dig massive areas using fewer workers. Petroleum refineries embodying U.S. $1 billion in investments continue to produce millions of gallons of oil derivatives a year with only a few hundred permanent workers. Emissions in these cases therefore may not follow cycles as closely. Our initial findings with a 200-year comparison of growth rates of the global economy and atmospheric carbon show that emissions first became correlated with and later diverged from economic cycles (Grimes and Roberts 1995).

A Series of Key Actors

World-system theory identifies a series of key actors. Political economy in general and world-system theory in particular point us to the importance of three sets of actors: states, capital, and labor (civil society) (Grimes 1988). The three have lately been called the "regulation triangle" by French theorists in the "regulation school" (e.g., Aglietta 1979; Lipietz 1986). In its search for the most important movers in the global system, world-system theory has isolated these actors in their manifestations as transnational corporations, core states, export (comprador) elites in the semiperiphery and periphery, and labor, most recently in popular organizations such as NGOs and unions.

This is a powerful tool for analyzing the sources and solutions to environmental outcomes, not a simple matter. Clearly, to understand the power of environmental movements to redirect local, national, or the global economy, we need to understand who are the prime directors of that economy. Although transnational corporations have to be a central focus of our analysis, it is also the case that smaller national firms will be more likely to pollute recklessly and evade emerging international environmental standards (Roberts 1996c). This is because their small size allows them to escape the attention of environmental activists. Their importance is highlighted by the statistics that small- and medium-sized firms are responsible for three-quarters of the global product (Strohl 1997). Be that as it may, capital—big or small—controls the majority of the sites of production globally and thereby has a massive influence on the degree of overall pollution and resource depletion.

States seek to control the activities of capital within their borders for two reasons: to retain popular legitimacy by appearing to implement popular will in the form of operating regulations and, equally importantly, to extract tax revenue. In the sixteenth to nineteenth centuries, states could and did expand to enclose within their boundaries the operations of most of the largest firms (e.g., the United States in particular, but also each of the colonial powers). Now, however, they can regulate only those corporate activities occurring inside their borders and even then only with an eye toward the effect of those regulations on the attractiveness to capital of competing states. Indeed, nation states may be becoming obsolete for environmental regulation because they are "too small for the big problems and too big for the small problems" (McMichael 1996). The complexity of supply chains in globalizing economies suggests that the "comprador" elite is changing and diversifying rapidly, scrambling to catch up with global diversification. Finally, both labor and environmental NGOs are themselves rapidly becoming linked globally, giving greater but highly conditional leverage to groups in the semiperiphery and periphery (Roberts 1996a; Barbosa 1993; Wapner 1996; Taylor 1995). Yet the future may witness a phase of global coalition-building that may yet constrain the freedom of global capital.

Peripheral Exploitation and Its Mechanism

Much world-system research has focused on peripheral exploitation and its mechanisms (coercion, extraction of economic surplus and resources). This topic has been touched upon already, but some additional aspects need exploration. The question of how two countries could enter into a relationship of allegedly "free trade" and one be enriched while the other impoverished was one of world-system theory's first issues. This question was central to the Dependency School thinkers from Latin America and Africa, and the question remains important today (see review in Shannon 1996).

Most of these writings traced the source of the surplus value from the periphery to the low rate at which labor is remunerated there (Amin 1974, 1976; Emmanuel 1972). Because of the oversupply of labor power in the periphery and the coercive political conditions, workers are paid little per unit of production. Only a few writers have considered that these differences in labor conditions might be themselves also dependent on the natural environment through some unequal "land rent" or "subsidy from nature" (Bunker, personal communication). Bunker looked to Volume 3 of *Capital* to examine Marx's interpretation of land rent. Meanwhile de Janvry's influential work *The Agrarian Question and Reformism in Latin America* (1981), while not explicitly world-system theory, did outline the importance of "disarticulated economies" (see below) and implied an impending contradiction in the long-term "mining the soil" by both plantation and *minifundia* agriculture alike.

One example is provided by Hecht, Anderson, and May (1988), who used the term "the subsidy from nature" to examine the use of naturally-occurring Babassú palms in Northeast Brazil to support poor seasonal laborers. The term deserves further exploration. We believe that several subsidies cheapen products from the world-system's non-core zones. First, urban informal (not fully proletarianized) labor and unpaid family workers (especially women) cheapen the cost of hiring workers and sustain them when they are not employed (Amin 1976; de Janvry 1981; Wolf 1982; Portes and Walton 1981; B. Roberts 1978). Second, several other irregularities push down living expenses in peripheral cities, such as squatter invaded land, micro-subsistence efforts, and so on (Roberts 1995b). Finally, rural subsistence enclaves may be declining as a percent of the periphery's population, but have remained important in absolute numbers.

If the "subsidy from the periphery's environment" (to butcher a phrase) is truly important for the functioning of capitalism—as is implied by theories of imperialism (see Harvey 1982)—the implications for the environment are massive. First, if the global economy requires the continuing incorporation of increasing parts of the globe into the economy, this implies peripheral nations are indeed key elements, and a destabilizing of their environment could destabilize the larger system. Estimates of soil erosion, for example, suggest that over half of the land in Central America and Africa is in a state of serious degradation (World Resources Institute 1993). What happens when these soils run out on a sufficiently large scale? The unequal exchange made possible by a "subsidy from the periphery's nature" theory suggests that the result will not be merely poverty, malnutrition, and refugees in those regions (as in the past), but the dramatic price rise of products from the periphery, as capital is compelled to internalize more and more of the full and true cost of peripheral labor. Thus it could be once again that the ecological contradiction is the ultimate one, a crucial point world-system theory has missed. We return to this point in the final section.

Structural Causes of Intra- and Inter-State Conflicts

World-system theory has attempted to ascertain the structural causes of intra- and inter-state conflicts. Wars have been shown to have devastating environmental consequences (Durham 1979). While environmentalists have pointed out this truism, little work has been done to link those environmental effects with a larger theory on the causes of wars. World-system theory has unearthed several structural causes for both the timing and motivation for war. That is, major global wars have been more common during certain phases of the "hegemonic cycle," when the central power of the declining hegemon is challenged by a rising new prospective power (Grimes 1988; Goldstein 1988; Chase-Dunn and Grimes 1995; Chase-Dunn 1989; Modelski and Thompson 1996). An example would be when one hegemonic nation (such as Holland, Spain, Great Britain, or the United States) loses its control over world trade and there is competition between would-be hegemons (e.g., Germany and Japan in the 1930s).

There has been an effort to tie these "hegemonic cycles" to the Kondratieff wave, but the results have as yet been inconclusive (Grimes 1988; Chase-Dunn and Grimes 1995; Modelski and Thompson 1996; Chase-Dunn and Pobodnik 1994; Shannon 1996). The general idea is that the Kondratieff heads into an upswing when a core country has the conditions allowing for a massive wave of investment in its productive infrastructure. Initially, this new infrastructure facilitates high profit activities incorporating the newest technologies. But, with time, this built environment lags ever further behind the march of innovation and also simply decays with use. Hence the rate of profit declines, and capital invests elsewhere. When it does, most vigorously at the beginning of the next Kondratieff, its location will be in yet another country, catapulting that country into position to rival and threaten the first country.

The 1969 "Soccer War" between Honduras and El Salvador has been tied directly to declining land quality and its shortage (Durham 1979; Faber 1993). Looking ahead, as resources become depleted and the land degraded into exhaustion (both most egregiously in the politically impotent periphery), global wars over hegemonic power will doubtless continue. However the spatial realm of production over which they will fight and preside must necessarily contract, as ever greater areas of land become unproductive waste. Meanwhile, local wars within the periphery over access to dwindling resources can be expected to increase.

Important policy implications derive from these points. An obvious one is that the role of the United Nations should be strengthened to enable more flexible military intervention that does not require United States approval so as to minimize the number, magnitude, and duration of wars in the periphery and thereby their environmental (as well as humanitarian) effects. A current example is provided by the war in Zaire, where the environmental damage is at least on a level with human costs. To summarize, by tying such wars to predictable cycles, world-system theory may provide a framework to understand past environmental degradation due to wars and a theoretical early warning system for anticipating future crises.

Socialism and the Transition to Capitalism

Finally and very briefly, world-system studies have paid great attention to socialism and the transition to capitalism. A series of authors and pundits have pointed out that the biggest environmental disasters in the past decades have in fact occurred in Eastern Europe and other state socialist countries around the world (e.g., Manser 1993; Harper 1996). What can world-system theory add to this discussion?

World-system theory can inform us on the nature of the transition from socialism, what state socialism was in the first place, and what the East-West competition meant to capitalism's unsustainability. First, world-system theory has argued repeatedly that these states were in fact capitalist in many ways, especially since they were competing in a capitalist world-system against capitalist states for political and economic power. Although labor was sometimes organized differently and economies were more influenced by state planning, state socialist societies, like Caribbean slave plantations in the past centuries, were "articulated" with global capitalism by trade in its produce, whether direct or indirect.

Concretely, world-system theory can inform predictions of whether and how environmental conditions might change in the ex-socialist states: who will tend to resist and who will support whatever improvements may occur. For example, the pessimistic but likely trajectory for Eastern Europe is for its role as a semiperiphery (already true in its "socialist" incarnation) to be ratified by investments from the West that would be typical of such investments in Latin America. Several East European countries are rapidly amassing foreign debts, similar to Latin American countries (World Bank 1995, 206–7). This suggests a sharpening of their subjugation to the terms of structural adjustment programs imposed by the World Bank and the International Monetary Fund. Should this prove to be the case, then we can anticipate the same situation now found in the Latin semiperiphery: continued employment of dated industrial technologies combined with coerced cheap labor generating massive air and water pollution. Certain showcase countries will be recipients of large amounts of concessionary-type aid such as grants in place of higher interest loans (see e.g., Stallings 1990).

One of the most intriguing questions is whether the logic of capitalism will be able to change now that there is no substantial anti-system with which it must compete. Can a new "Global Commonwealth" (Wagar 1996) or governance structure emerge out of the need to deal with increasingly global problems of which environmental issues are primary? To pose the negative expectation, will capitalism find a new enemy to replace the Soviets, such as Muslims or death itself as medicine becomes our new Cold War spending sink? As O'Connor (1973) suggested long ago, was the *warfare* state necessary to support the welfare state? We return to these provocative questions in the concluding section.

MERGING ENVIRONMENTAL SOCIOLOGY AND WORLD-SYSTEM THEORY: PROSPECTS, ISSUES, AND A RESEARCH AGENDA

Rather than being theoretical purists, we take the omnivorous approach and remind readers that world-system theory is one type of comparative political economy which needs the insights of other parts of that field and those of many others. On environmental issues, world-system theory must work hard to incorporate the decades of work in human geography, international studies, and environmental sociology. That may be the easy part: more contentious will be the suggestion that the field examine work in environmental psychology and gender and cultural studies.

However we believe that world-system theory has certain advantages over other approaches as an integrative framework for understanding environmental damage in global capitalism. Obviously world-system theory brings improvements to our understanding over modernization approaches and stage theories of development and underdevelopment. We find particularly dangerous the implications of development theorists that believe poor countries will follow the paths of development taken by today's wealthy countries and their (Kuznets' curve) propositions that conditions will first get worse and then reverse themselves at some turning-point. Beyond these obvious points we believe that world-system theory can increase our understanding of a series of environmental issues: global, transnational, those related to exports, material cycling, those which have tended to oscillate cyclically, those which are distributed unequally in the global hierarchy of poor and rich nations, and so on. That is, we believe world-system theory has a potential contribution to make on most pressing environmental issues, such as global warming, deforestation, resource depletion, water struggles, ozone, food crops, the increasing frequency of major storms and human recovery from them, biodiversity, compensation of locals for management of neighboring endangered species, enactment and enforcement of international treaties, and coordination of NGOs (Grimes 1997).

It is important to acknowledge the issues which have remained underplayed or entirely unaddressed in world-system theory and which will have to be addressed to "Green" world-system theory or applied more widely to environmental sociological understanding. The weaknesses with world-system theory are well summarized by Shannon (1996, 213ff.). He points out, for example, that world-system theory tends to overemphasize economic explanations, while remaining virtually mum on culture. This old materialism debate discussed above goes back to Marx and Weber, and certainly we will not resolve it here. However, despite our argument for the importance of material explanations, we believe that nonmaterial motivations, such as meeting culture-specific and evolving consumption expectations and more simply the desires for status, power, love, jealousy, fun, ego-gratification, etc., do have critical environmental implications, as do other social structural elements that cannot be "read off" a society's means of material production and reproduction. However we would propose that attention to these types of causation be combined with attention to a society's material system of survival.

We agree with Shannon that world-system theory explanations are often *overdetermined* externally to nations. World-system theory has been weak on its analyses of culture and individual agency. Imprecision and poorly operationalized concepts (Shannon 1996, chap. 6) are major shortcomings which this young field must address. Shannon makes the important point that zones of the world system—core, semiperiphery, and periphery—do violence to the diversity of nations (Shannon 1996, 213). This is an important point and in our own work we have chosen continuum measures as discussed above. However the question remains of whether any one index of WSP can capture a stratification system which is multidimensional. Finally, Shannon levels tough critiques that gender still has not been adequately addressed in the field, that world-system theory's arguments are often teleological, and that the historical accounts are often overgeneralized in the search for meta-narratives (as is true of any global theory of social change or development).

Each of these weaknesses is reparable given suitable attention, and none is fatal to the central insights of the paradigm. They can be fixed. Despite these shortcomings, we believe it clear that world-system theory carries with it vitally important environmental implications. While we regret that these implications have yet to be more thoroughly pursued, we hope that our work, along with others working on similar lines, will accelerate the employment of the tools of world-system theory to the growing environmental crisis.

NOTES

1. Should this turn out to be the case, the consignment of capitalism to the dustbin of history would be the same ecological degradation that has likewise trashed all prior modes of accumulation (e.g., Sanderson 1995; Chew 1995).

2. Sanderson (1985) makes the excellent point that while the opportunities of ruling classes are constrained by macro structures (the world economy in which they operate), sometimes they choose effectively and other times not. The wisdom of their decisions is

determined by micro issues internal to the political and historical situation in the nation itself.

3. The term "globalization"—that the network of world trade and politics is growing ever tighter—is itself an "old" world-system theory concept which has become a buzzword for this entire decade or phase of capitalism, now used far beyond world-system theory circles (Chase-Dunn and Grimes 1995).

4. This relationship is not necessary, since automation also depends on the price of labor and the culturally variable desire of capitalists to gain total control over the production process.

REFERENCES

Aglietta, M. 1979. *A Theory of Capitalist Regulation.* London: New Left Books.

Amin, Samir. 1974. *Accumulation on a World Scale.* 2 vols. New York: Monthly Review Press.

Amin, Samir. 1976. *Unequal Development.* New York: Monthly Review Press.

Arrighi, Giovanni, and Jessica Drangel. 1986. The Stratification of the World-Economy: An Exploration of the Semi-Peripheral Zone. *Review* 10(1): 9–74.

Ayres, Robert U. 1989. Industrial Metabolism and Global Change. *ISSJ: International Social Science Journal* 41: 363–73.

Barbosa, Luis. 1993. The World-System and the Destruction of the Brazilian Rain Forest. *Review* 16(2): 215–40.

Barbosa, Luis. 1996. The People of the Forest Against International Capitalism: Systemic and Anti-Systemic Forces in the Battle for the Preservation of the Brazilian Amazon Rainforest. *Sociological Perspectives* 39(2): 317–31.

Barkin, David. 1995. Wealth, Poverty, and Sustainable Development. Working Paper. Lincoln Institute.

Barnham, Bradford, Stephen G. Bunker, and Denis O'Hearn. 1994. *States, Firms, and Raw Materials: The World Economy and Ecology of Aluminum.* Madison: University of Wisconsin Press.

Braudel, Fernand. 1972. *The Mediterranean and the Mediterranean World in the Age of Philip II.* 2 vols. New York: Harper & Row.

Braudel, Fernand. 1981. *The Structures of Everyday Life.* Vol. 1 of *Civilization and Capitalism, 15th–18th Century.* New York: Harper & Row.

Braudel, Fernand. 1982. *The Wheels of Commerce.* Vol. 2 of *Civilization and Capitalism, 15th–18th Century.* New York: Harper & Row.

Braudel, Fernand. 1984. *The Perspectives of the World.* Vol. 3 of *Civilization and Capitalism, 15th–18th Century.* New York: Harper and Row.

Braverman, Harry. 1974. *Labor and Monopoly Capital.* New York: Monthly Review Press.

Brenner, R. 1977. The Origins of Capitalist Development: A Critique of the Neo-Smithian Marxism. *New Left Review* 104: 25–92.

Brewer, Anthony. 1980. *Marxist Theories of Imperialism: A Critical Survey.* London: Routledge &Kegan Paul.

Bunker, Stephen G. 1985. *Underdeveloping the Amazon: Extraction, Unequal Exchange, and the Failure of the Modern State.* Urbana: University of Illinois Press.

Bunker, Stephen G. 1989. The Eternal Conquest. *NACLA Report on the Americas* 23(1): 27–36.

Bunker, Stephen G. 1994. Flimsy Joint Ventures in Fragile Environments. In *States, Firms, and Raw Materials: The World Economy and Ecology of Aluminum,* ed. Bradford Barnham, Stephen G. Bunker, and Denis O'Hearn, 261–96. Madison: University of Wisconsin Press.

Burbach, Roger, and Patricia Flynn. 1980. *Agribusiness in the Americas.* New York: NACLA and Monthly Review Press.

Buttel, Frederick H. 1978. Social Structure and Energy Efficiency: A Preliminary Cross-National Analysis. *Human Ecology* 6(2): 145–64.

Buttel, Frederick H. 1987. New Directions in Environmental Sociology. *Annual Review of Sociology* 13: 465–88.

Chase-Dunn, Christopher. 1989. *Global Formation: Structures of the World-Economy.* Cambridge, MA: Basil Blackwell.

Chase-Dunn, Christopher, and Bruce Pobodnik. 1994. The World-System and World State Formation. Paper presented at the Thirteenth World Congress of Sociology of the Internatonal Sociology Association, Symposium II, Bielefeld, Germany, July 22.

Chase-Dunn, Christopher, and Peter Grimes. 1995. World-Systems Analysis. *Annual Review of Sociology* 21: 387–417.

Chase-Dunn, Christopher, and Thomas D. Hall, eds. 1991. *Core/Periphery Relations in Precapitalist Worlds.* Boulder, CO: Westview Press.

Chase-Dunn, Christopher, and Thomas D. Hall. 1996. Ecological Degradation and the Evolution of World-Systems. Paper presented at the American Sociological Association session on The Environment and the World Economy, New York, August 17.

Chew, Sing C. 1995. Environmental Transformations: Accumulation, Ecological Crisis, and Social Movements. In *A New World Order? Global Transformations in the Late Twentieth Century,* ed. David A. Smith and József Borocz, 201–15. Westport, CT: Praeger.

Chew, Sing C. 1997. For Nature: Deep Greening World-Systems Analysis for the 21st Century. *Journal of World-Systems Research* 3(3): 381–402. http://csf.colorado.edu/wsystems/jwsr.html.

Chilcote, Ronald H. 1984. *Theories of Development and Underdevelopment.* Boulder, CO: Westview Press.

Ciccantell, Paul. 1994. The Raw Materials Route to the Semiperiphery: Raw Materials, State Development Policies and Mobility in the Capitalist World-System. American Sociological Association Annual Meetings, Los Angeles, August.

Covello, Vincent T., and R. Scott Frey. 1990. Technology-Based Environmental Health Risks in Developing Nations. *Technological Forecasting and Social Change* 37: 159–79.

de Janvry, Alain. 1981. *The Agrarian Question and Reformism in Latin America.* Baltimore: Johns Hopkins University Press.

Dicken, Peter. 1992. *Global Shift: The Internationalization of Economic Activity.* New York: Guilford Press.

Dunlap, Riley E., and Willam R. Catton, Jr. 1994. Struggling with Human Exemptionalism: The Rise, Decline, and Revitalization of Environmental Sociology. *The American Sociologist* 25: 5–30.

Durham, W. 1979. *Scarcity and Survival in Central America: The Ecological Origins of the Soccer War.* Stanford, CA: Stanford University Press.

Emmanuel, Arghiri. 1972. *Unequal Exchange: A Study of the Imperialism of Trade.* New York: Monthly Review Press.

Evans, Peter. 1979. *Dependent Development: The Alliance of Multinational, State, and Local Capital in Brazil.* Princeton, NJ: Princeton University Press.

Evans, Peter B., Dietrich Rueschemeyer, and Theda Skocpol, eds. 1985. *Bringing the State Back In.* New York: Cambridge University Press.

Faber, Daniel. 1993. *Environment under Fire: Imperialism and the Ecological Crisis in Central America.* New York: Monthly Review Press.

Foster, John Bellamy. 1994. *The Vulnerable Planet: A Short Economic History of the Environment.* New York: Cornerstone Books.

Frank, Andre Gunder. 1969. The Development of Underdevelopment. *Monthly Review* 18(4): 17-31.

Frank, Andre Gunder. 1978. *World Accumulation 1492–1789.* New York: Monthly Review Press.

Frank, Andre Gunder, and Barry K. Gills, eds. 1993. *The World System: Five Hundred Years or Five Thousand?* New York: Routledge.

Frank, Andre Gunder, and Barry K. Gills. 1991. 5000 Years of World System History: The Cumulation of Accumulation. In *Core/Periphery Relations in Precapitalist Worlds,* ed. C. Chase-Dunn and T. Hall, 67–112. Boulder, CO: Westview Press.

Freudenburg, William R. 1981. Women and Men in an Energy Boomtown: Adjustment, Alienation, and Adaptation. *Rural Sociology* 46(2): 220–44.

Frey, R. Scott. 1993a. The Capitalist World Economy, Toxic Waste Dumping, and Health Risks in the Third World. Paper presented at the American Sociological Association Annual Meeting. Miami, FL.

Frey, R. Scott. 1993b. The International Hazardous Waste Trade. Paper presented at the American Sociological Association Annual Meeting. Miami, FL.

Gellert, Paul K. 1996. Concentrating Capital with a Spatially Diffuse Commodity: The Political Ecology and Economy of the Indonesian Timber Industry. Paper presented at the American Sociological Association Annual Meeting. New York, August 16–20.

Gereffi, Gary, and Miguel Korzeniewicz, eds. 1994. *Commodity Chains and Global Capitalism.* Westport, CT: Praeger.

Goldstein, Joshua. 1988. *Long Cycles: Prosperity and War in the Modern Age.* New Haven, CT: Yale University Press.

Gould, Kenneth. 1994. Transnational Trade Deregulation: A Collision Course with Sustainable Development? St. Lawrence University, Department of Sociology. Paper presented at the Conference on the Politics of Sustainable Development. University of Crete, Greece, October 21–23.

Gramling, Robert, and Sarah Brabant. 1990. The Impact of a Boom/Bust Economy on Women's Employment. Paper presented at the annual conference, American Sociological Association, Washington, DC, August.

Grimes, Peter E. 1988. Long Cycles, International Mobility, and Class Struggle in the World-System. Dissertation proposal.

Grimes, Peter E. 1996. Economic Cycles and International Mobility in the World-System: 1790–1990. Ph.D. diss., Johns Hopkins University.

Grimes, Peter E. 1997. The Horsemen in the Killing Fields. Paper presented at the 22nd Meeting of the Political Economy of the World-System, Santa Cruz, April 3–5.

Grimes, Peter E., J. Timmons Roberts, and Jodie Manale. 1993. A World-Systems Analysis of Deforestation. Paper presented at the American Sociological Association Annual Meeting.

Grimes, Peter E., and J. Timmons Roberts. 1995. Oscillations in Atmospheric Carbon Dioxide and Long Cycles of Production in the World Economy, 1790–1990. Paper presented at the American Sociological Association, Annual Meeting, Washington, DC, August.

Grimes, Peter E., and J. Timmons Roberts. 1996. Shifting Correlates of National Carbon Efficiency in the World System. Paper presented at the American Sociological Association Annual Meeting. New York, August 16–20.

Grossman, Gene M., and Alan B. Krueger. 1995. Economic Growth and the Environment. *Quarterly Journal of Economics* 110 (May): 353–77.

Harper, Charles L. 1996. *Environment and Society: Human Perspectives on Environmental Issues.* Upper Saddle River, NJ: Prentice-Hall.

Harvey, David. 1982. *The Limits to Capital.* Chicago: University of Chicago Press.

Hecht, Susanna B., Anthony B. Anderson, and Peter May. 1988. The Subsidy from Nature: Shifting Cultivation, Successional Palm Forests, and Rural Development. *Human Organization* 47 (Spring): 25–35.

Howard, Michael C., ed. 1993. *Asia's Environmental Crisis*. Boulder, CO: Westview Press.

Inglehart, Ronald. 1995. Political Support for Environmental Protection: Objective Problems and Subjective Values in 43 Societies. *PS: Political Science and Politics* 23(1): 57–72.

Kazis, Richard, and Richard L. Grossman. 1991. *Fear at Work: Job Blackmail, Labor and the Environment.* Philadelphia: New Society Publishers.

Kloppenburg, Jack R. 1988. *First the Seed: The Political Economy of Plant Biotechnology.* New York: Cambridge University Press.

Laclau, Ernesto. 1971. Feudalism and Capitalism in Latin America. *New Left Review* 67 (May–June).

Lenin, Vladimir I. 1939. *Imperialism: The Highest Stage of Capitalism*. New York: International Publishers.

Lipietz, Alain. 1986. Beyond Global Fordism? *New Left Review* 132: 33–47.

Logan, John R., and Harvey L. Molotch. 1987. *Urban Fortunes: The Political Economy of Place*. Berkeley: University of California Press.

Mann, Michael. 1986. *A History of Power from the Beginning to A.D. 1760.* Vol. 1 of *The Sources of Social Power.* New York: Cambridge University Press.

Manser, Roger. 1993. *Failed Transitions: The Eastern European Economy and Environment since the Fall of Communism.* New York: New Press.

Marx, Karl. 1967. *Capital.* 3 Vols. New York: International House.

Mazur, Allan, and Eugene Rosa. 1974. Energy and Life-Style. *Science* 186: 607–10

McMichael, Philip. 1996. Globalization: Myths and Realities. *Rural Sociology* 61(19): 25–55.

Modelski, George, and William R. Thompson. 1996. *Leading Sectors and World Powers: the Coevolution of Global Politics and Economics*. Columbia: University of South Carolina Press.

O'Connor, James. 1973. *The Fiscal Crisis of the State.* New York: St. Martin's Press.

O'Connor, James. 1989. Capitalism, Nature, Socialism: A Theoretical Introduction. *Capitalism, Nature, Socialism* 1(1): 11–38.

Perelman, Michael. 1977. *Farming for Profit in a Hungry World: Capital and the Crisis in Agriculture.* New York: Universe Books.

Portes, Alejandro, and John Walton. 1981. *Labor, Class, and the International System.* Orlando: Academic Press.

Ranis, Gustav. 1990. Contrasts in the Political Economy of Development Policy. In *Manufacturing Miracles: Paths of Industrialization in Latin America and East Asia,* ed. Gary Gereffi and Donald L. Wyman, 207–30. Princeton, NJ: Princeton University Press.

Reed, David, ed. 1992. *Structural Adjustment and the Environment.* Boulder, CO: Westview Press.

Roberts, Bryan. 1978. *City of Peasants: The Political Economy of Urbanization in the Third World.* London: Edward Arnold.

Roberts, J. Timmons. 1992. Forging Development, Fragmenting Labor: Subcontracting and Local Response in an Amazon Boomtown. Ph.D. diss., Johns Hopkins University.

Roberts, J. Timmons. 1994. Economic Crisis and Environmental Policy [Brazil]. *Hemisphere* 6(1): 26–30.

Roberts, J. Timmons. 1995a. Trickling-Down and Scrambling-Up: Informal Sectors and Local Benefits of a Mining "Growth Pole" in the Brazilian Amazon. *World Development* 23(3): 385–400.

Roberts, J. Timmons. 1995b. Subcontracting and the Omitted Social Impacts of Development Projects: Household Survival at the Carajás Mines in the Brazilian Amazon. *Economic Development and Cultural Change* 43(4): 735–58.

Roberts, J. Timmons. 1996a. Predicting Participation in Environmental Treaties: A World-System Analysis. *Sociological Inquiry* 66(1): 38–57.

Roberts, J. Timmons. 1996b. Global Restructuring and the Environment in Latin America. In *Latin America in the World Economy,* ed. Roberto P. Korzeniewicz and William C. Smith. Westport, CT: Greenwood.

Roberts, J. Timmons. 1996c. Green Labels and Greenwashing: Corporate Environmental Campaigns and International Environmental Standards. Paper presented at the Mid-South Sociological Association, Little Rock, AK, October.

Roberts, J. Timmons, and F. Nai-Amoo Dodoo. 1995. Population Growth, Sex Ratio and Women's Work on the Contemporary Amazon Frontier. *Yearbook of the Conference of Latin American Geographers,* 91–105.

Roberts, J. Timmons, and Peter E. Grimes. 1997. Carbon Intensity and Economic Development 1962–1991: A Brief Exploration of the Environmental Kuznets Curve. *World Development* 25(Feb.): 2.

Rudel, Thomas K. 1989. Population, Development, and Tropical Deforestation: A Cross-National Study. *Rural Sociology* 54: 327–38.

Sanderson, Stephen. 1988. *Macrosociology: An Introduction to Human Societies.* New York: Harper & Row.

Sanderson, Stephen K. 1995. *Social Transformations: A General Theory of Historical Development.* London: Basil Blackwell.

Santos, Roberto Araujo de Oliveira. 1980. *Historia Econômica da Amazônia 1800–1920.* São Paulo: T. A. Queiroz.

Schafer, D. Michael. 1994. *Winners and Losers: How Sectors Shape the Developmental Prospects of States.* Ithaca, NY: Cornell University Press.

Schnaiberg, Allan. 1980. *The Environment: From Surplus to Scarcity.* New York: Oxford University Press.

Shannon, Thomas R. 1996. *An Introduction to the World-System Perspective,* 2nd ed. Boulder, CO: Westview Press.

Simonis, Udo E. 1989. Ecological Modernization of Industrial Society: Three Strategic Elements. *ISSJ: International Social Science Journal* 41: 347–61.

Skocpol, Theda. 1985. Bringing the State Back In: Strategies of Analysis in Current Research. In *Bringing the State Back In,* ed. P.Evans, D. Rueschemeyer, and T. Skocpol, 3–42. New York: Cambridge University Press.

Smith, David A. 1994. Uneven Development and the Environment: Toward a World-System Perspective. *Humboldt Journal of Social Relations* 20(1): 151–75.

Smith, David A., and Douglas R. White. 1992. Structure and Dynamics of the Global Economy: Network Analysis of International Trade 1965–1980. *Social Forces* 70(4): 857–93.

Stallings, Barbara. 1990. The Role of Foreign Capital and Economic Development. In *Manufacturing Miracles: Paths of Industrialization in Latin America and East Asia,* ed. Gary Gereffi and Donald Wyman, 267–91. Princeton, NJ: Princeton University Press.

Strohl, Derek. 1997. Research on Multinational Corporations and Large and Small Businesses in Mexico: Hypotheses on Comparative Levels of Pollution. Mimeo.

Taylor, Bron Raymond, ed. 1995. *Ecological Resistance Movements: The Global Emergence of Radical and Popular Environmentalism.* Albany: State University of New York Press.

Terlouw, Cornelis Peter. 1992. *The Regional Geography of the World-System.* Utrect: Koninklijk Nederlands Aardrijkskundig Genootschap.

Tucker, Richard P., and John F. Richards, eds. 1983. *Global Deforestation and the Nineteenth-Century World Economy.* Vol 1. Durham, NC: Duke University Press.

Wagar, W. Warren. 1996. Socialism, Nationalism, and Ecocide. *Review* 19(3): 319–33.

Wallerstein, Immanuel. 1974. *The Modern World-System I: Capitalist Agriculture and the Origins of the European World Economy in the Sixteenth Century.* Vol. 1. New York: Academic Press.

Wallerstein, Immanuel. 1979. *The Capitalist World-Economy.* New York: Cambridge University Press.

Wallerstein, Immanuel. 1989. *The Modern World System III: The Second Era of Great Expansion of the Capitalist World-Economy, 1730–1840s.* New York: Academic Press.

Wapner, Paul K. 1996. *Environmental Activism and World Civic Politics.* Albany: State University of New York Press.

Wolf, Eric. 1982. *Europe and the People without History.* Berkeley: University of California Press.

Wood, Charles H., and José Alberto Magno de Carvalho. 1988. *The Demography of Inequality in Brazil.* Cambridge: Cambridge University Press.

World Bank. 1992. *World Development Report 1992.* New York: Oxford University Press.

World Bank. 1995. *World Development Report 1995.* New York: Oxford University Press.

World Resources Institute. 1993. *World Resources.* New York: Oxford University Press.

Wright, Angus. 1990. *The Death of Ramón González: The Modern Agricultural Dilemma.* Austin: University of Texas Press.

Yearley, Steven. 1996. *Sociology, Environmentalism, Globalization: Reinventing the Globe.* London: SAGE Publications.

Part II

5

Ecological Relations and the Decline of Civilizations in the Bronze Age World-System: Mesopotamia & Harappa 2500 B.C.–1700 B.C.

Sing C. Chew

INTRODUCTION[1]

Our understanding of long-term change from a world-systems perspective has focused on the social, economic, and political relations underlining the dynamics of the world-system. These relations, for analytical purposes, have covered two main areas: a) the dynamics of the accumulation process circumscribed by the global division of labor and punctuated by long economic cycles of stagnation and expansion, and b) the competitive rivalry between core states for global market share and hegemony. For at least the last five thousand years of world history, the human activities surrounding the world accumulation process have resulted in ecological degradation and crisis (Chew 1997a). In spite of this historical degradation of nature, the overall focus of world-systems analysis to date has been to pinpoint the anthropogenic factors to account for the dynamics and logics of capital accumulation over the long term.What also warrants a place in our understanding of world history is an examination of our relations and collisions with nature in the human quest to reproduce the material basis of life. These other dimensions of human history, as McNeill (1990, 20–21) observes, "also deserve a place in any really satisfactory account of the past; they, too, ought to be woven into the narrative of the rise and elaboration of separate civilizations and cultures and viewed as ecumenical processes comparable in importance with the rise of a world system of economic complementarity and cultural symbiosis."

At this stage of world history, a reorientation by world-systems analysis to address the nature-culture nexus, for example, in looking at ecological relations surrounding human organization and reproduction, is important. There is a need to realize that ecological relations, natural processes, and climatological rhythms impact on and, on occasions, condition the reproduction and expansionary capacities of societal systems. This chapter introduces a discussion of these considerations by analyzing the political, economic, and ecological relations circumscribing the dynamics of the reproduction and decline of two socioeconomic organizations (Mesopotamia and Harappa)[2] during the Bronze Age. It suggests that ecological relations are as primary as the political-economic dimension in our understanding of social change over the long term (Chew 1997a, 1997b). In fact, what we find over world history is that ecological and political–economic relations intertwine and condition the growth and demise of human communities.

BRONZE AGE WORLD-SYSTEM: THE CASE OF MESOPOTAMIA AND HARAPPA

Accumulation: Trade, Production, and Core-Hinterland Relations

Recent discussions on world accumulation within world-systems analysis have suggested the continuity of the systemic process of the accumulation of capital over the long term (Frank 1993; Frank and Gills 1992). Capital, in this case, is by no means what some Marxists would define as wage-labor capital, but is used in a more general context to mean the accumulation of surplus dependent on the state of technology and sociocultural practices of the period in question. Along this vein, production, trade, and commodity exchanges underly the dynamics of the process of accumulation. In addition to bulk goods, preciosities are also considered part and parcel of the capital accumulation process in the ancient world.[3] Therefore, in order to account for ecological degradation and the demise of socioeconomic organizations, we need first to articulate the dynamics, extent, and connectivity of production, and trade exchanges in the Bronze Age world system. From this, we can then understand the scope of ecological degradation that ensues.

According to Childe (1957, 3), the regular use of copper and bronze in production and reproduction of human communities suggests the existence of an organized international trade during the Bronze Age. Childe's logic rests on the fact that the locations of the centers of urbanization/accumulation (such as in Egypt, Mesopotamia, and Harappa) were in the fertile alluvial valleys and plains devoid of metallic deposits. Therefore, to meet the economic need of large quantities of metals for production, other regions and areas had to supply the centers of urbanization/accumulation with the necessary metals. Though Childe did not explore this system of production, trade relations, and linkages in a detailed systematic fashion for Mesopotamia and Harappa, anthropological and archaeological studies have outlined an established structure of production, trade relations, and exchange occurring during the Bronze Age (i.e., Asthana 1982; Possehl 1982; Oppenheim 1979; Bag 1985; Tosi 1982; Gadd 1932; Lamberg-Karlovsky 1975; Ratnagar 1981, 1991, 1994; Edens 1992; Kohl 1987; Allen 1992; Piggott 1950; Allchin and Allchin 1982; Moorey 1994). What emerges from these studies is a series of connections occurring between these two far-flung urban centers and also with their surrounding hinterlands and the utilization of the land and the seas to execute these transactions. By no means were these exchanges via trade alone; at various times, military expeditions and wars furthered the search for natural resources that the core centers lacked and/or had already exhausted by unsustainable exploitation. Thus we have the development of a structure of linkages that is determined by ecological conditions, trade routes, merchant communities, and political relations. It suggests, as well, that these linkages were circumscribed within a structure of core-periphery relations characterized by a division of labor, with Mesopotamia and the Harappan being the core centers of accumulation (Edens 1992, 121; Kohl 1987). During this time period of the third millennium, Mesopotamia and the Harappan were not the only core centers; Egypt and the Namazga civilization of southern Central Asia were also part of the core. This seems to suggest that there were several economic/political trade structures that often overlapped with multiple core centers and associated peripheries/hinterlands in the Bronze Age.

The above economic structural relations were in place by 2500 B.C., when we witness a flow of preciosities and bulk goods circumscribing the developed urban centers from southern Mesopotamia to the northwestern Indus Valley. This structure of economic and political linkages covered the plains of Elam, including the Zagros ranges, the upper and central Persian Gulf, the Oman peninsula (especially the southeastern part), Persian and Pakistani Baluchistan, and the Indo-Gangetic divide. Notwithstanding certain specific ecological characteristics that existed in the river valleys, mountainous ranges, and desert conditions of this geographic expanse, the area has a common ecological feature comprising "an arid or semi-arid climate with summer droughts and an unevenly distributed and unreliable rainfall in the winter months amounting to less than 250 mm per annum" (Ratnagar 1981, xvi). In addition to natural resource extraction, this common arid ecological characteristic therefore conditioned the type of socioeconomic practices being undertaken, such as winter agriculture and the herding of sheep and goats. In riverine valleys, the natural potential for high agricultural productivity could be seen, especially in Mesopotamia and the Indus (Harappa), along with the feasibility of communication/exchanges via land, river, and sea. This potential was further increased by the use of irrigation. Depending on the region, the mountain ranges, plateaus, and plains provided the wood and other natural resources such as metals, marble, gypsum, and alabaster for the urbanized communities in the riverine cities of the respective core centers in Mesopotamia and Harappa. The latter, in turn, specialized in agricultural production and the manufacture of pottery, oils, wool, leather items, etc., along with wood, metal, and stone objects manufactured from imported materials for exchange. Against the expanse of this structure of economic linkages, there existed a division of labor conditioned by the ecological surroundings and the political-economic structures. The high productivity in agriculture, along with the specialization in manufactured products (such as textiles) facilitated by slave labor, enabled these urban centers with their associated social structures to maintain their political-economic dominance in this economic system.

Mesopotamia, Elam, and the Gulf

If one travels from west to east in this Bronze Age economic system of exchange, there are different trading points and cultural communities reflecting various levels of development and urbanization (see Map 5.1). The urbanized communities of Mesopotamia traded in essentially three directions, utilizing both land and water. The river Euphrates served the trade routes to the west and northwest (such as Syria and Anatolia), linking the eastern parts of the Mediterranean with the Gulf. To the east and northeast, the land transport linked Mesopotamia to the Iranian plateau and beyond. The southern trade routes were connected via the Gulf to Dilmun, Magan, and Meluhha.

In the third millennium B.C., southern Mesopotamia was composed of urbanized communities with the principal port cities of Ur, Lagash, and Susa being the main entry points for goods from the east and southeast of this political-economic system (Edens 1992; Oppenheim 1979). Canals (not only for irrigation) and natural waterways were built to facilitate the economic exchange. Large-scale urban communities with population sizes ranging from 10,000 to more than 50,000 were in existence and included palaces and temples (Redman 1978). All this urbanization meant the intensive utilization of natural resources of the immediate surroundings and also the importation of some natural resources such as timber from distant reaches (Indus Valley) of the economic system. High-quality timber from the Zagros mountains, the Caspian area, western Asia, eastern Mediterranean, Punjab, and the Indus Valley were obtained via military expeditions and trade (Willcox 1992). The ecological surroundings of southern Mesopotamia, especially its rivers, provided not only a plentiful and reliable water supply but also deposits of alluvium during annual floods. Such conditions, buttressed by an extensive irrigation system of canals, dams, and reservoirs, provided high surpluses of cereals produced at low prices that were used as a basis for trade exchanges with other regions east and southeast of Mesopotamia that were agriculturally poorer (Leemans 1960, 115; Moorey 1994). In addition to cereals, oils, wool, and leather products, expensive items such as garments, textiles, and metallic objects (including preciosities) manufactured from imported materials were also exported (Adams 1981). The raw materials for the latter items were obtained at relatively low prices from other parts of the economic system. It was a case of "unequal exchange" (Allen 1992; Kohl 1978; Algaze 1989). Imports from eastern and southeastern parts of the system such as metals, timber, and other items such as ivory, pearls, conch, and beads satisfied elite and mass consumption. Silver, obtained through trade (from Elam and east-central Anatolia), plunder, or tribute, was one of the capital sources used to pay for the imports (Leemans 1960, 132). Such a scale of exports and imports suggests that the goods and materials exported and imported were both luxuries and necessities, with the latter targeted for mass consumption. In some cases, a luxury became a necessity once it became integrated into the political economy of consumption. The exports from Mesopotamia were usually shipped in high-prowed ships built of long giant beams tied together and coated with natural bitumen, with sails made of reed mats. These ships would range as far as Meluhha, located near the mouth of the Indus River (Diakonoff 1991a).

The scale of economic production (textiles, oils, leather, and other objects) with the high generation of agricultural surplus in Mesopotamia facilitated the exploitation of other areas within this economic system. For these hinterland areas

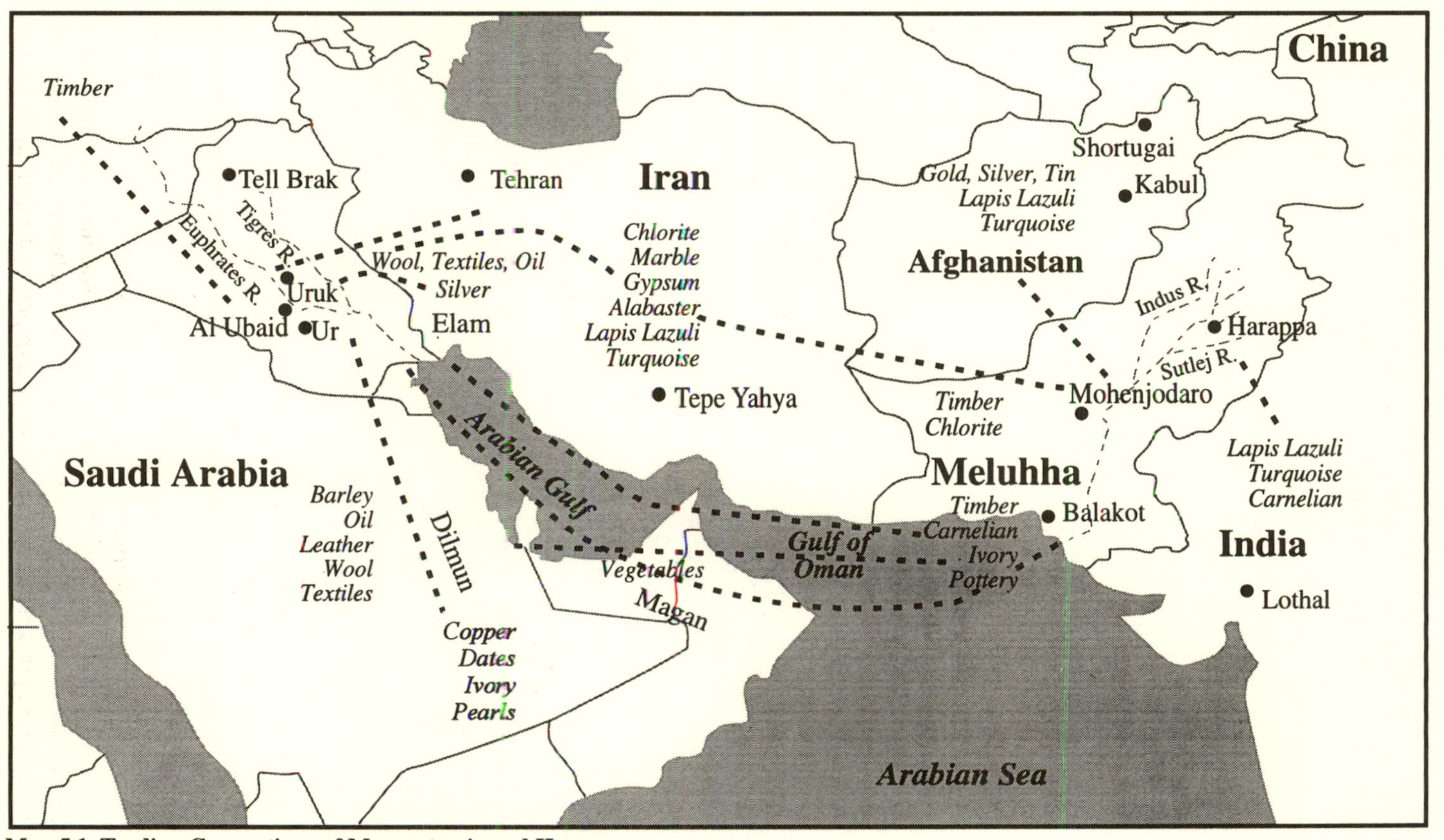

Map 5.1 Trading Connections of Mesopotamia and Harappa

required Mesopotamian products, such as oils and grains, not only for elite consumption but also for mass consumption (Lamberg-Karlovsky 1975). For example, manufactured goods from Mesopotamia were exchanged for chlorite found at Tepe Yahya on the Iranian plateau. Yahya artisans worked at a set production rate and were paid with goods by middlemen merchants. In return, these middlemen with the local Yahya elites (as demonstrated by the collection of high-priced items found in some third millennium tombs in the area) benefited from the high profit of the chlorite sold to high-demand centers in Mesopotamia (Lamberg-Karlovsky 1975). Out of this arose a division of labor whereby the outlying regions/hinterland participated in economic exchange by concentrating on a particular productive activity conditioned wholly by the available natural resources. Elsewhere to the southeast of southern Mesopotamia, grains such as barley, along with wool and textiles, provided the exchange products with lands such as Dilmun.

This form of economic imperialism was also coupled with political expansion/imperialism, starting as early as the Uruk period (4000 B.C.) (Algaze 1993; Diakonoff 1991b). Outposts were established in the hinterland areas to control crucial nodes of trading routes along the Syro-Mesopotamian plains, and they were established among native polities controlling communication and trade in the northern plains of the Zagros. Such imperialistic ventures can be divided into conjunctures (Lamberg-Karlovsky 1986). Starting in 3300 B.C., the Sumerians with their centralized administration undertook a series of colonizing missions on their periphery. This conjuncture represented an initial attempt of the city-states of Mesopotamia to colonize the less bureaucratically integrated societies of the region. It was an expansionist policy that lasted for at least two centuries (Lamberg-Karlovsky 1986). The second conjuncture, starting around 2900 B.C., saw the Elamites, a regional competitor to the Sumerians, pursuing an imperialist-expansionist policy in colonizing regions of the Iranian plateau. During the final conjuncture from 2400 B.C. onwards, expansionist Mesopotamian policies linked the Gulf and beyond.

Northeast of this economic system, beyond the Mesopotamian plains, are the Zagros ranges with their bountiful resources in timber, marble, gypsum, alabaster, limestone, and other metals. East and south of the Zagros lies Elam, which was endowed with bountiful natural resources such as timber, precious metals, and stone. In Elam, Mesopotamian merchants exchanged cattle, precious metals, oils, and timber for shipbuilding and for wool, barley, tin, and silver. These relations were also punctuated with frequent military expeditions and wars, a consequence of the Mesopotamian search for raw materials (timber, stone, metals). The Elamites sought to capture the wealth (copper, gold, slaves, etc.) of Mesopotamia.[4]

To the east of Elam on the Iranian plateau were various communities (Tepe Yahya, Sialk, and so on) that produced the valuable metals and stones consumed by the core center. In the previous pages, we have already discussed the economic imperialism between southern Mesopotamia and Tepe Yahya.

A main trading nexus of this economic system was the Gulf. Sumerian and Akkadian documents underline the extensive trade between Mesopotamia and Dilmun and Magan. Dilmun appears to be an entrepôt for the Gulf trade for goods (timber, copper, ivory, etc.) coming from the East as far away as Meluhha and Harappa and from western centers such as Egypt, Mesopotamia, and the Mediterranean. Magan also had extensive trade relations with Mesopotamia. These relations lasted until 2000 B.C. when Dilmun seems to have replaced Magan in terms of trade relations in the southeastern part of the economic system stretching from Mesopotamia to the Indus Valley. Magan, like Dilmun, was also one of the entrepôt sites on the Gulf. Large quantities of barley, garments, wool, leather objects, and oil from Mesopotamia were exchanged for copper, precious stones, carnelian, ivory, and vegetables from Magan. The carnelian and ivory were transshipment products from as far away as the Indus Valley.

The growth in the Gulf trade reached a peak between 2000 and 1750 B.C. and thus enhanced the local elites in Dilmun and Magan who controlled this trade in terms of sourcing and producing goods for exchange with Mesopotamia. The increasing urbanization in these places meant that the imported cereals amplified the socioeconomic position of the local elites as they controlled the grain imports. The trade decline started after 1750 B.C. and coincided with the decline of the other core center located to the easternmost part of this economic system: Harappa. For Mesopotamia, it seems that the trading volume never revived, and trading activities shifted away from Dilmun and Magan and became concentrated more to the north and west with the importation of the needed natural resources such as copper, precious stones, and timber.

Harappa

As we push further eastwards of Mesopotamia and Elam, the predominance of the other core center in trade relations, Harappa, can be seen. Bordering the western edge of Harappa were Kulli communities located along all the major avenues of communication of Makran. Lying to the west of the Indus Valley, this area, especially the hills between Liari and Bela, provided the metals such as copper for Harappan manufacturing and trade (Ratnagar 1981). Harappan weights and seals have been found at Kulli sites, and the influence of the Harrapans on the Kullis can be seen in the great similarity in pottery styles between the Harrapans and Kullis. It is also likely that the Kullis acted as merchant middlemen between the Harappans and the Mesopotamians (Piggott 1950). Therefore, the Kulli region formed part of the vast hinterland that supplied the natural resources to Harappa for its production processes. In this regard, Ratnagar (1991) has even suggested that Kulli sites represent client chiefdoms.

We find a parallel to the urbanization and accumulation processes of Mesopotamia in the eastern section of this overarching system in the other core center, Harappa. The extent of the Harappan influence/authority reached as far west as the modern border between Pakistan and Iran, extended as far north as the foothills of the Himalayas, and stretched southwards along the west coast of India as far as the Gulf of Cambay. As in Mesopotamia, we witness again the natural potential for food production in the alluvial plains where a large number of Harappan sites have been excavated. The urban centers were located in the river valleys of the Indus and its tributaries. The Harappan influence spread over 1.3 million square kilometers (Agrawal and Sood 1982). Its two main cities of Harappa and Mohenjo-daro were each over 120 hectares in area, with populations of 37,155 and 41,000, respectively, around 2500 B.C. (Ratnagar 1981; Allchin and Allchin 1982; Bag 1985; Marshall 1931). The degree of urbanization and architectural development can be seen in the spatial design of the cities and towns and the architectural contours of the buildings and homes. Drainage and sewer systems were integrated into the urban design. Some of the cities (Mohenjo-daro, Harappa, Ganweriwala, Banawali, Kalibangan, Mitathal, Rakhi Garhi, Sutkagen-dor, Desalpur, Kotara Juni Karan, Dholavira, Surkotada, and Lothal) were surrounded by walls with a citadel towering over the urban complex (Lal 1993). The large cities contained palaces and granaries (measuring 15 x 6 meters) to hold many hundreds of metric tons of grain. One Great Bath was also found at Mohenjo-daro measuring 12 x 7 meters and nearly 3 meters deep. Burnt or fired bricks were used in enormous quantities for building construction along with sun-dried ones. Large amounts of fuel resources such as wood were utilized in the manufacture of such bricks (Allchin and Allchin 1982; Piggott 1950; Lal 1993; Wheeler 1968). In addition, timber was also used for the flat roofs of the houses, and beams with spans as much as 4 meters were utilized. The buildings and houses were connected by a sewage system with many houses having settling pits for waste water, which was directed out of the city by brick-lined channels. Every street of the urban areas was supplied with one or two drains made with burnt bricks, fine gypsum, and sand cement (Bag 1985). This level of urban transformation thus suggests a high consumption of natural resources from the immediate surroundings and from the hinterland areas that sustained the continued development and reproduction of urban life.

The region in which the Harappan communities were located had the same arid zone characteristics found in other parts of this vast economic system which extends to Mesopotamia. Barley, wheat, cotton, and sesame were grown on the rich alluvium soil of the river valleys. Gravity-flow irrigation employing the shaduf lift was used in agricultural cultivation (Leshnik 1973). We can assume that there was surplus production as the urban centers contained central granaries for storage and perhaps redistribution (Piggott 1950).[5] Items manufactured for consumption and trade included textiles, beads, pottery, copper/bronze objects including axes, knives, spearheads, jewelry, toys, carvings, seals, and weights. These were made in the urban centers, while specific urban areas (lower towns) were designated for industrial production (Kenoyer 1997). The raw materials for these industries were chiefly copper, steatite, agate, and carnelian. The heating of the carnelian, copper casting, and steatite glazing must have required a high volume of fuel that was obtained from the immediate surroundings or from the hinterland. For example, the heating of steatite for hardening purposes required temperature ranges from 900–1000 degrees centrigrade (Kenoyer 1997).

The diversity of economic production and trade underlines the varied linkages that the Harappans sought, exchanged, and dominated within this region of the economic system of the Bronze Age. Harappan linkages extended to Baluchistan, Afghanistan, Iran, Central Asia, peninsular India, and the lands bordering the Persian Gulf right to southern Mesopotamia (Asthana 1982; Kenoyer 1997). Semi-precious stones such as lapis lazuli and turquoise were obtained from Afghanistan, Iran, peninsular India, and Central Asia for export to areas as far away as southern Mesopotamia. Besides Afghanistan and Oman, northwestern India (Rajasthan and Gujerat) also provided copper to the Harappans (Asthana 1982). The Harappans incorporated Gujerat into its political and economic sphere so as to ensure the constant flow of natural resources. Like the Mesopotamians, the Harappans would develop communities/outposts that would handle the exchange and process the natural resources obtained. Carnelian, which was used in the manufacture of beads and pendants, was sought for in peninsular India where we also find Harappan sites. Timber, a major export of the Harappans to Mesopotamia utilized for building and home construction, was sought for in the Western Ghats, the Jammu Ranges, and the Panjab piedmont. Collection centers on the western part of Gujerat were established to facilitate the flow of timber for consumption in the Indus Valley and for export via Meluhha and Harappan coastal ports.

It appears that the trade and manufacturing mechanisms of the Harappans were highly evolved in terms of the mechanics of ensuring a constant supply of natural resources and a market for the manufactured products in the hinterland and in other parts of the economic system of the Bronze Age.[6] Gateway settlements or outposts were embedded in a wide periphery to facilitate the flow of goods and natural resources (Asthana 1982; Algaze 1993; Possehl and Raval 1989). These settlements were established at locations near strategic trade routes/passes (e.g., Nausharo) or located in hinterland areas close to the natural resources/commodities (e.g., Shortugai), or near coastal areas to facilitate the maritime trade (e.g., Lothal). These settlements, some of which were fortified (such as Sutkagen-dor and Sutka-koh), provided the access points for the flow of natural resources much needed by the manufacturing economy of the Harappans. In addition, there were trading posts for the exchange of Harappan manufactured goods and some agricultural products. Some settlements, such as Lothal, not only were procurement centers for natural resources but also had manufacturing and modification activities (Possehl and Raval 1989). Thus, manufacturing was not located only at Harappan cities in the Indus Valley but also in these established hinterland gateway settlements. In this light, the Harappan towns had also a divison of labor in the overall trade and manufacturing activities. Some towns were connected with different kinds of production activities. For example, Harappa specialized in metal tool manufacturing, while Mohenjo-daro concentrated on metal objects and textiles.

Such were the trading and production mechanisms circumscribing the exchange between the Harappans and their vast hinterland. Besides these center-hinterland exchanges, there were also center-center exchanges during the third and second millennia B.C. As we have indicated earlier, Harappan goods were exchanged throughout the Bronze Age economic system as far west as southern Mesopotamia. Dales (1977) has periodized that there was a shift in trading routes of the Bronze Age economic system away from the Central Asian land trade routes to a maritime route by the mid-third millennium B.C. connecting the Indus Valley via the Gulf to southern Mesopotamia. The trade on this maritime route started to decline between Mesopotamia, the Gulf, and the Indus Valley by the mid-second millennium. From Harappa, copper, timber, conch shells, ivory, various stones, carnelian beads, and pigments were exported. Some of these products that the Harappans had obtained from their hinterland were then re-exported. From Mesopotamia, it was bulk goods and utilitarian items such as food stuffs and textiles, shipped through the Gulf.

ECOLOGICAL RELATIONS: MESOPOTAMIA AND HARAPPA

Reproducing the materialist needs of the urbanized and stratified communities of Mesopotamia and Harappa required an intensive and transformative relationship with nature. Urbanization essentially transformed the nature of material culture, which in turn undoubtedly led to an exchange of manufactured products for raw materials. A whole range of new materials and products, including manufacturing processes, came into being to meet these transformed material needs. They ranged from agricultural implements, eating equipment, furniture, textiles, transport equipment, ornaments, and even weapons of war. What arose also were manufactories, workshops, and other industrial structures to meet the needs of urbanized communities.

The set of ecological relations that resulted was not restricted to the immediate surroundings of these urbanized communities but was extended to their hinterland areas. As a result, the resource needs of the core centers (such as Mesopotamia and Harappa) and the resultant consequences such as ecological degradation and transformation of the landscape were also transmitted ("exported") throughout the other areas of the system. This economic linkage in terms of resource utilization and trade does not mean that the appearance of ecological crisis (as a consequence of intensive utilization of nature) in one part of the economic system (for example, Mesopotamia) would translate to the appearance of ecological crisis in other parts (Dilmun, Harappa, Magan, etc.) of the system, though ecological degradation could be an outcome. Instead, economic crises of supply and demand could arise in these other parts as a consequence of the division of labor and trade linkages. Therefore, an ecological crisis in Mesopotamia could mean the lowering of agricultural output or production and consequently a reduction in the overall supply of goods to other parts of the system. Concomitantly, such reductions would impact other regions such as Dilmun and Harappan through a diminished demand for their materials and goods. This suggests that ecological and political-economic relations are intertwined in conditioning the transformation of regions of the economic system.

Furthermore, besides considering ecological relations as having an impact on the dynamics of the Bronze Age economic system that encompassed Mesopotamia and Harappa, climatological changes (in terms of rainfall and temperature fluctuations) and geological shifts could also impact on the reproductive capacities of these urbanized communities. In the long run, such natural changes did impact the politico-economic positions of Mesopotamia and Harappan within the system. Therefore, a consideration of all these factors and conditions will help us to better understand the demise of these two urban complexes.

Agricultural Production and Urbanization

As already noted, third millennium B.C. Mesopotamia concentrated on agricultural production as well as manufacturing. Devoid of natural resources such as copper, tin, stones, ivory, and timber for its manufacturing processes, all these materials had to be imported from various points throughout the Bronze Age economic system, stretching from Lebanon in the north to as far east as the Harappan. Urbanization was also a feature of the transformation of the landscape during this period. The relationship between culture and nature is underlined by the economic activities of agriculture, manufacturing, and urbanization.

Agriculture was undertaken with the construction of an irrigation system. Initially, simple gravity fed irrigation systems were utilized. With irrigation, fresh regions were opened up for highly productive agricultural exploitation as early as the fifth millennium B.C. Control of the flooding of the Euphrates and the Tigris was done through the construction of dams and channels. Extensive resources were used to further agricultural production. For example, the main irrigation canals were lined with burnt bricks and the joints sealed with asphalt. By 1800 B.C. the irrigation system had expanded to about 10,000 square miles (Carter and Dale 1974). Crops included several varieties of grain and other plant species for oil production and textile manufacture. This activity thus transformed the landscape.

The extensive use of irrigation required that the canals remained clear. However, Mesopotamia suffered from siltation problems clogging up its irrigation canals. The cause of this siltation had social roots in the extensive deforestation of the northern forests to meet Mesopotamia's fuel needs for manufacturing (copper smelting), urban social consumption, and building materials such as roof beams, posts, rods, planks, and boards (Willcox 1992; Rowton 1967). Agricultural implements such as hoes, plough-shares, and sickle handles were also made out of wood, and wood was required for shipbuilding. Overgrazing of the land via animal husbandry further exacerbated the siltation problem. With deforestation and overgrazing, erosion proceeded to strip the topsoil, with the latter pouring into the streams and rivers and being carried hundreds of miles downstream into canals and channels. The history of Mesopotamian agriculture is one replete with the struggle of having to keep the irrigation canals clear of silt. When neglected as a consequence of social turmoil, it impacted on agricultural production.

A further threat to Mesopotamian agriculture was the fact that the water of the rivers descending from Anatolia and the Zagros contained a high proportion of dissolved salts which deposited into the alluvium (Jacobsen and Adams 1958; Oates and Oates 1976). The intense summer heat with temperatures of over 120 degrees (F) and the very flat land resulted in poor drainage. This, in turn, brought the salt to the surface of the fields which in the long run had a major impact on agricultural productivity, as the sodium ions tend to be absorbed by colloidal clay particles leaving the resultant structure impermeable to water (Oates and Oates 1976; Hughes 1975; Jacobsen and Adams 1958). High salt concentrations obstruct germination and also impede plants from absorbing water and nutrients. This history of intensive utilization of natural resources (such as wood) impacted on the agricultural economic productivity and thus determined the trajectory of socioeconomic transformation. The exploitation was not only within the immediate hinterland (up to Lebanon) but extended as far east as the Harappan. The effects of the degradation of the land and exploitation of natural resources by the Mesopotamians were also transferred to other parts of the economic system.

The growth and consumptive pattern of the Harappan civilization paralleled that occurring in Mesopotamia. The Harappans depended on their hinterland for natural resources to meet their internal urban needs, for exports to their hinterland and also to Dilmun and Mesopotamia (Possehl and Raval 1989). Like Mesopotamia, the Harappans specialized in agriculture, manufacturing, and processing of goods and natural resources. Unlike Mesopotamia, besides the utilization of various woods for home and building construction, the Harappans' urban communities relied on tremendous amounts of burnt bricks for construction (Hughes 1975; Piggott 1950; Ratnagar 1981; Allchin and Allchin 1982; Wheeler 1968). Burnt bricks in standard sizes measuring 11 x 5.5 x 2.5 inches to 24 x 16 x 4 inches were found in Mohenjo-daro, Harappa, and Dabar Kot (Piggott 1950). An estimate of 5,000,000 bricks constituted the visible site of Mohenjo-daro (Fairservis 1979b) Such large-scale utilization of bricks for buildings, sewer systems, and canals meant that large quantities of wood were required. This intensive utilization of a natural resource was exacerbated by the export of wood to the Gulf. Besides the intensive utilization of wood for construction and for export, wood was also in heavy demand for the manufacture of beads, copper, and steatite glazing. Fueling this resource need were the immediate surroundings of the Indus Valley and the hinterland areas, such as the northeastern Punjab (on the Siwaliks and the foothills) and the Western Ghats. As a consequence, there was extensive deforestation in the Indus Valley and the identified hinterlands (Hughes 1975; Piggott 1950; Ratnagar 1981, 1986; Wheeler 1968; Allchin and Allchin 1982). Combined with overgrazing, the end result of this widespread deforestation included desiccation, flooding, and erosion.

Monoculture led to intensive utilization of the land and cattle grazing further exacerbated ecological degradation. With an increase in population, further pressure was added to increase land utilization. Fairservis (1979b, 87) has commented that "the growth of population, human and animal, dependent solely on a rabi crop, created seasonal stresses which in the end caused the abandonment of most of the region" of Sind. Ecological degradation of the landscape was widespread and increased in intensity as the millennium progressed. The continued socioeconomic growth of the Harappan civilization coupled with the growth in population pushed the carrying capacity of the environment to its limits.

Ecological degradation affected not only the urbanized communities but also the biodiversity of the area. The climatic conditions of most of the area of the Harappan civilization also made a suitable habitat for the rhinoceros, elephant, water buffalo, crocodile, and tiger (Wheeler 1968; Piggott 1950). These species are depicted in the stamped seals that have been unearthed. However, none of these species seem to have survived in later periods with the exception of the tiger. The loss of these species was a result of human activities (Possehl 1997).

THE DEMISE

The ecological relations of the two core urban complexes (Mesopotamia and Harappa) indicate an intensive exploitation of nature to meet not only domestic needs but also commercial requirements in other parts of the Bronze Age economic system. Furthermore, they also underscore a drive to control and exploit nature which is reflected throughout the history of urbanized communities for at least the last 5,000 years (Chew 1997a). Exploitative ecological relations upset the balance in the natural environment, exposing such complexes to ecological stresses, natural climatic changes, and tectonic shifts. This vulnerability, coupled with a decline in trade, civil/social conflicts, and competitive state rivalries, engendered the conditions for the demise of these urban complexes.

As stated previously, the climatic zone in which the Harappan civilization extended was an arid one. Agrawal and Sood (1982), reviewing climatological data, have suggested that between 3000 B.C. and 1500 B.C. the weather was wetter than usual in this zone. This period was followed by a severe aridity until 1 A.D. Ratnagar (1981) points as well to this climatic shift and has further suggested that a small scale oscillation to drier conditions occurred between 1800 B.C. and 1500 B.C. This latter arid period is crucial, because it was during this time that the Harappan civilization proceeded to decline. Especially for arid regions, minor shifts in terms of wetness (notwithstanding the major shift that occurred post-1500 B.C.) would spell severe ecological stress. Thus, starting from the second millennium B.C., the increasing aridity would have placed great stress on the Harappan civilization. The shift in rainfall pattern, coupled with increasing salinity, would have placed severe stress on its agricultural productivity.

The tectonic shifts that diverted water courses, which in turn transformed some rivers into dry river beds, further exacerbated the aridity and thus affected the socioeconomic life. Agrawal and Sood (1982) note tectonic shifts that diverted the course of the Satluz and the easterly rivers away from the Ghaggar, which gradually died into a lakelike depression during this period. The Ghaggar was alive until the late Harappan Period (1800 B.C.) but by the time of the Painted Grey Ware period (1000 B.C.) it had turned dry. Thus, in northern and western Rajasthan, unstable river systems impacted on socioeconomic life. Furthermore, tectonic disturbances also cut off Lothal from its feeder river and eventually the port's access to the sea.

This thesis of tectonic shifts impacting on the reproduction of socioeconomic life has also been advanced to account for the demise of the Harappans, especially Mohenjo-daro and its associated communities (Raikes 1964; Raikes and Dales 1977; Dales 1979; Sahni 1956). The thesis is that a tectonic uplift generated a dam about 90 miles downstream from Mohenjo-daro in an area near Sehwan, which caused the normal discharge of the Indus to accumulate in a growing reservoir that over time caused the flooding of Mohenjo-daro. Mud-brick platforms were erected of burnt bricks to keep the city safe from the flooding waters. Embankments about 70 feet wide and over 25 feet high have been excavated. Archaeological excavations have indicated silt accumulation up to a vertical distance of 70 feet sandwiched between successive levels of occupations of Mohenjo-daro. The silt deposits are not of the type spread by normal flooding of fast flowing rivers but seem to be of the kind similar to still-water conditions.[7] Dales (1979), building on this, has suggested that enmeshed in this deterioriating condition, the weakened state was unable to send help to its inhabitants in the northern frontier when they were threatened by tribal incursions.

Beyond the climatological changes and tectonic shifts arguments for the demise of the Harappans, it has been suggested that overcultivation, overgrazing, salinity, deforestation, and flooding are factors generating the decline of the Harappan urban complex (Fairservis 1979a; Raikes and Dyson 1961; Wheeler 1968). The flooding of the Indus would have had the effect of increasing the water table for a considerable distance on both sides of the channel. With the salt content of the water table being high, and the mean temperature about 90 degrees, poor leaching led over time to large areas becoming unfit for cultivation. Along with the wearing out of the landscape through overcultivation, overgrazing, and deforestation, this would have generated severe stress on socioeconomic reproduction. It has led Fairservis (1979b, 88) to suggest that "the evidence again demonstrates a failure to come to grips fully with what must have become an increasingly acute situation—the destruction of the local ecological patterns and the consequent failure of food resources."

The identification of the above conditions as possible factors leading to the decline of the Harappans around 1700 B.C. furthers the need to consider ecological relations, and natural climatological and tectonic shifts as important variables impacting on the reproduction of the Harappan civilization.[8] This consideration needs to be coupled with other anthropogenic factors such as the decay of administrative structures, social conflicts, and border threats in the north and west (Dyson 1982; Piggott 1950). Notwithstanding all these anthropogenic and ecological factors that are *specific* to geographic parameters of the Harappan civilization, we need to consider the economic linkages the Harappans had with other parts of the Bronze Age economic system that also facilitated the demise of this urban complex.

The decline of the Harappans coincided with that of southern Mesopotamia. Trade between the Harappans and Mesopotamia and the Gulf region, as we have indicated, flourished from 2500 B.C. until 1750 B.C. After this, the Gulf trade disappeared (Leemans 1960).[9] The end of this flourishing maritime trade coincided with the demise (1700–1500 B.C.) of the Harappan civilization. The urbanized communities of the Harappan civilization, including the infrastructure and surrounding communities, had come to specialize in the manufacture of products and natural resources for export. Thus, when its exports to the Gulf and beyond disappeared, it could no longer reproduce the accumulation process that had sustained its urban growth. This led to outmigration to the rural areas of the north and south. With the accumulation process slowing down, the outlying trading towns and outposts in the hinterland, devoid of prosperous support from the Harappan core, gradually merged with the countryside. Coupled with the problems of (normal and abnormal) flooding and the deterioriating ecological conditions, the decline proceeded unchecked.

The contemporaneous decline of two core centers of accumulation underline the systemic connectedness of the Bronze Age economic system. Harappan exports were impacted by the slowdown in the Gulf trade and declining political eminence of the southern cities of Mesopotamia where the exports entered. This decline also had its roots in the ecological relations that existed along with climatological changes in terms of temperature and rainfall.

Between 3100 and 1200 B.C., the Near East experienced a drop in precipitation, and we find that wetter conditions did not return until 850 B.C. (Neumann and Sigrist 1978). This increase in aridity was followed by much warmer weather during the winter months. The latter increase in temperature, especially between 1800 and 1650 B.C., has been confirmed for Mesopotamia in terms of harvest dates for barley during this period in comparison to a later period after 850 B.C. (Neumann and Sigrist 1978). These climatological changes, especially an increase in temperature, must have been accompanied by a rise in evapotranspiration. For irrigated agriculture this would have meant a demand for more water. The enhanced application of irrigated water had a deleterious effect on agricultural lands that possessed a salinity problem. Such conditions were prevalent in southern Mesopotamia between 2400 B.C. and 1700 B.C. and led to a crisis in agricultural productivity (Jacobsen and Adams 1958; Adams 1981).

This crisis in agricultural productivity has to be understood within a wider context of political-economic relations and the ecological relations of Mesopotamia. No doubt climatological changes such as warmer temperatures and decreasing rainfall are serious factors relating to agricultural productivity, but such shifts might not have so much impact if Mesopotamia had attempted to maintain a balance with nature. As stated previously, this was not the case. The stratified society pursued intensive socioeconomic activities to produce a surplus for domestic consumption as well as for exports (in the form of grains and woolen textiles) to the Gulf and beyond. The scale of intensity required extensive deforestation, maximal utilization of agriculture, and animal husbandry. Furthermore, with a state structure requiring tax payments, etc., the farmers were forced to produce an increasing surplus to meet the reproductive needs of the system (Jacobsen and Adams 1958).

Besides requiring surplus production, the Third Dynasty of Ur (2150–2000 B.C.) also concentrated on certain economic activities such as the production of wool and the development of a large scale textile industry. This further pushed the need to increase agricultural productivity in the form of feed grains such as barley for the sheep (Adams 1981). Population increases and state initiatives to establish new towns, peopled by conquered populations for the purpose of pursuing agricultural and textile manufacturing, added to the range of economic practices that required heightened resource utilization (Gelb 1973). The end result of these political and economic initiatives was an intensification of agricultural production that pushed the agricultural lands to the limit.

As indicated previously, salinization was a problem in Mesopotamia from the third millennium onwards. Wheat and barley that were grown in the mid-fourth millennium in equal portions had shifted to more barley cultivation by the end of the third millennium. With the salinity problem, after 1700 B.C., the cultivation of wheat was completely given up as barley was a more salt-tolerant grain. In addition, barley was also the preeminent feed for sheep, whose end product, wool, was a commodity required for long-distance trade. However, this change in grain cultivation did not solve the issue of soil salinity as is reflected in the harvest yields. Between 3000 and 2350 B.C. (Early Dynastic Period), crop yields were 2,030 liters per hectare. By 2150–2000 B.C. (Third Dynasty of Ur), the yield has fallen to 1,134 liters per hectare. By 1700 B.C., crop yields has slipped to 718 liters per hectare (Jacobsen and Adams 1958). The agricultural productivity crisis reached its nadir when cultivable land was kept in production with yields of only 370 liters per hectare (Adams 1981). With such yields, the "burden on the cultivator had become a crushing one" (Adams 1981, 152).

To meet this agricultural crisis, seeding rates were increased. Between 2150 and 2000 B.C., 55.5 liters of seed were planted per hectare. This volume was double that of the previous period between 3000 and 2350 B.C. Furthermore, without considering the repercussions, alternate-year fallowing was also violated in order to maintain the maximum amount of land available for cultivation (Gibson 1970). Fallowing is the traditional method of handling salinization. In Mesopotamia, by leaving the land to fallow, wild plants such as *shok* and *agul* drew moisture from the water table and dried up the subsoil. This prevented the water from rising and bringing the salts to the surface. When the land was cultivated again, the dryness of the subsoil allowed the irrigation water to leach salt from the surface and drained it below the root level. Fallowing therefore returned the land to its cultivation potential. Reducing or violating fallow times endangered productivity of the land.

Southern Mesopotamia never recovered from the disastrous decline in agricultural yields that accompanied the salinization process. As a consequence, the cultural and political centers of power shifted from the south to the north, with the rise of Babylon in eighteenth century B.C. This demise of southern Mesopotamia and the resultant transfer of power to the north as a consequence of ecological relations and climatological changes needs to be connected as well to the collapse of the Third Dynasty of Ur as a consequence of the activities of tribal groups such as the Amorites and the coalitions that resulted (Yoffee 1988). When combined, these anthropogenic and ecological explanations enrich our understanding of the vectors that led to the demise of southern Mesopotamia and the consequent rise of Babylon. Furthermore, with the power base shifting northwards, the trade in metals and timber with sources in Syria and Anatolia increased. The central Mesopotamian cities thus relied on this northern hinterland more for its resources than the southern cities did. Such developments led to the strengthening of the land trading routes of the north and northeast and the demise of the maritime trade routes of the Gulf. The shift engendered the disappearance of the Gulf trade, and the repercussions were felt throughout the trading points along the Gulf up to the Indus Valley.

CONCLUSION

This analysis suggests that the contemporaneous demise of Harappa and southern Mesopotamia was a consequence of their economic connectedness in the Bronze Age economic system and their comparable ecological relations with the natural environment. The character of these ecological relations is circumscribed by the dynamics of the process of unequal accumulation and the pace of urbanization that existed then. In addition to these two processes (unequal accumulation and urbanization), there are others that engender ecologically exploitative relations that most often lead to ecological stress. Ultimately, the continued ability to sustain unequal accumulation and urbanization while maintaining a successful balance with the natural environment turns into crisis during a conjuncture when climatological changes (perhaps even tectonic shifts), ecological stress, civil strife, state rivalry, border incursions, and wars punctuate the historical moment. No doubt a greater degree of technological development would have provided breathing room to overcome these crises. Further incorporation of other areas to enhance the depleted resources would have been another avenue of handling the crisis. In the long run, however, such balancing acts would reach asymptotic limits; only a transformation of the historical system (in terms of the logic of accumulation) would overcome the exploitative ecological and social relations. In this regard, our efforts to understand the demise of a historical system need to focus not only on the anthropogenic, but also on the ecological limits circumscribing the process of capital accumulation.

NOTES

1. Many thanks to Diana Axelsen, Gunder Frank, Bill Devall, George Modelski, Shereen Ratnagar, and Andy Szasz for their comments on an earlier version of this chapter. Support for this research was from grants provided by the California State University System and the Humboldt State University Foundation.

2. The use of the word Harappa/Harappan is not to designate specifically the city of Harappa in the Indus valley during the Bronze Age. It is used to represent the urbanized communities located in Bronze Age Indus Valley that are known historically as forming the Harappan Civilization.

3. Unlike Wallerstein (1974), Schneider (1977) has argued that preciosities did generate systemic effects in premodern times.

4. The paucity of archaeological evidence makes its difficult to assess Elam's position within this economic system. The type of resources and materials that it had for exchange suggest that its economy was based on resource extraction. This would indicate that it was not as economically transformed as southern Mesopotamia and therefore would not necessarily have been a part of the core centers of accumulation during the third millennium. Ratnagar (1994) has suggested that Elam, at times, was politically annexed to Mesopotamia during the third millennium. Lamberg-Karlovsky (1986) however, has a different view of Elam's standing to Mesopotamia during certain periods of the third millennium. Elam achieved political unity by the third millennium where there were three ruling houses: the kings of Awan, Simashki, and the "Grand Regents." Textual sources showed that Awan had a position of supremacy over Mesopotamia around 2250 B.C. However, by 2325 B.C., Sargon of Akkad conquered Elam.

5. Ratnagar (1986) has questioned the agricultural prosperity of the Harappan region vis-à-vis southern Mesopotamia of this time period, in light of the ecological conditions requiring lift irrigation and the density of urbanization.

6. Our position differs from Shaffer(1982) who has indicated that the Harappan civilization was primarily dependent on internal exchange.

7. This thesis has been questioned by Lambrick (1967).

8. The dating for the decline of the Harappan civilization varies according to different Harappan scholars. Wheeler(1968) has suggested the earliest decline dates from 3250 B.C.–2750 B.C. while Gadd (1932) has pinpointed 2350 B.C.–1700 B.C. for the period of decline based on his analysis of the seals. Ratnagar (1981) has identified the period between 2000 B.C.–1760 B.C. for the phase of decline while Possehl and Raval (1989) have suggested the phase ending later around 1000 B.C. Fairservis' (1979b) dating of 1750 B.C. for the end seems to encompass these differences and it also dovetails with the decline of the Gulf trade which the Harappans participated actively in.

9. Frank (1993) has indicated an overall economic decline (B-phase) from 1800/1750–1600/1500 B.C. during this phase of the Bronze Age.

REFERENCES

Adams, Robert McC. 1981. *Heartland of the Cities.* Chicago: University of Chicago Press.

Agrawal, D. P. and R. K. Sood. 1982. Ecological Factors and the Harappan Civilization. In *Harappan Civilization,* ed. G. Possehl, 223–31. New Delhi: Oxford University Press.

Algaze, Guillermo. 1989. The Uruk Expansion: Cross-Cultural Exchange in Early Mesopotamia. *Current Anthropology* 30(5): 571–608.

———. 1993. Expansionary Dynamics of Some Early Pristine States. *American Anthropologist* 95(2): 304–33.

Allchin, Bridget and Raymond Allchin. 1982. *The Rise of Civilization in India and Pakistan.* New York: Cambridge University Press.

Allen, Mitchell. 1992. The Mechanisms of Underdevelopment: An Ancient Mesopotamian Example. *Review* 25(3): 453–76.

Asthana, Shashi. 1982. Harappan Trade in Metals and Minerals: A Regional Approach. In *Harrapan Civilization,* ed. G. Possehl, 271–85. New Delhi: Oxford University Press.

Bag, A. K. 1985. *Science and Civilization in India.* New Delhi: Navrang.

Carter, Vernon, and Tom Dale. 1974. *Topsoil and Civilization.* Norman: University of Oklahoma Press.

Chew, Sing C. 1997a. Accumulation, Deforestation, and World Ecological Degradation 2500 B.C. to A.D.1990. In *Advances in Human Ecology,* Vol. 6. Westport, CT: JAI Press.

———. 1997b. For Nature: Deep Greening World-Systems Analysis for the 21st Century. *Journal of World-Systems Research* 3(3): 381–402. (http//csf.Colorado.edu/wsystems/jwsr.html).

Childe, Gordon. 1957. The Bronze Age. *Past and Present* 12: 2–15.

Dales, G. F. 1977. Shifting Trade Patterns between Iranian Plateau and the Indus Valley. In *Le Plateau Iranien et l'Asie Centrale,* ed. J. Deshayes, 67–78. Paris: CNRS.

———. 1979. The Decline of the Harappans. In *Ancient Cities of the Indus,* ed. G. Possehl, 307–12. New Delhi: Vikas.

Diakonoff, I. M. 1991a. The City States of Sumer. In *Early Antiquity,* ed. I. M. Diakonoff, 67–83. Chicago: University of Chicago Press.

———. 1991b. Early Despotisms in Mesopotamia. In *Early Antiquity,* ed. I. M. Diakonoff, 84–97. Chicago: University of Chicago Press.

Dyson, Robert. 1982. Paradigm Changes in the Study of the Indus Civilization. In *Harappan Civilization,* ed. G. Possehl, 572–81. New Delhi: Oxford University Press.

Edens, C. 1992. The Dynamics of Trade in Ancient Mesopotamian World-System. *American Anthropologist* 94: 118–39.

Fairservis, Walter. 1979a. The Harappan Civilization: New Evidence and More Theory. In *Ancient Cities of the Indus,* ed. G. Possehl, 50–65. New Delhi: Vikas.

———. 1979b. The Origin, Character, and Decline of an Early Civilization. In *Ancient Cities of the Indus,* ed. G. Possehl, 66–89. New Delhi: Vikas.

Frank, A. G. 1993. Bronze Age World System Cycles. *Current Anthropology* 34(4): 383–429.

Frank, A. G., and Gills, B. K. 1992. The Five Thousand Year World System: An Interdisciplinary Introduction. *Humboldt Journal of Social Relations* 18(2): 1–80.

Gadd, C. J. 1932. Seals of Ancient Indian Style Found at Ur. *Proceedings of the British Academy* 18: 3–22.

Gelb, I. J. 1973. Prisoners of War in Early Mesopotamia. *Journal of Near Eastern Studies* 32: 70–98.

Gibson, McGuire. 1970. Violation of Fallow and Engineered Disaster in Mesopotamian Civilization. In *Irrigation's Impact on Society,* ed. T. Downing and M. Gibson. Tucson: University of Arizona Press.

Hughes, Donald J. 1975. *Ecology in Ancient Civilizations.* Albuquerque: University of New Mexico Press.

Jacobsen, T., and R. M. Adams. 1958. Salt and Silt in Ancient Mesopotamian Agriculture. *Science* 128: 1251–58.

Kenoyer, J. M. 1997. Trade and Technology of the Indus Valley: New Insights from Harappa, Pakistan. *World Archaeology* 29(2): 262–80.

Kohl, Philip. 1978. The Balance of Trade in Southwestern Asia in the Mid-Third Millennium. *Current Anthropology* 193: 480–81.

———. 1987. The Ancient Economy, Transferable Technologies, and the Bronze Age World System: A View from the Northeastern Frontiers of the Ancient Near East. In *Centre and Periphery in the Ancient World,* ed. M. Rowlands et al., 13–24. Cambridge, UK: Cambridge University Press.

Lal, B. B. 1993. A Glimpse of the Social Stratification and Political Set-up of the Indus Civilization. *Harappan Studies,* ed. G. Possehl and M. Tosi, 63–71, vol. 1. New Delhi: Oxford University Press.

Lamberg-Karlovsky, C. C. 1975. Third Millennium Exchange and Production. In *Ancient Civilization and Trade,* ed. Jeremy Sabloff and C. C. Lamberg-Karlovsky, 341–68. Albuquerque: University of New Mexico Press.

———. 1986. Third Millennium Structure and Process: From the Euphrates to the Indus and the Oxus to the Indian Ocean. *Oriens Antiquvvs* 25: 189–219.

Lambrick, H. T. 1967. The Indus Flood Plain and the "Indus Civilization." *The Geographical Journal* 133(4): 483–89.

Leemans, W. F. 1960. *Foreign Trade in the Old Babylonian Period as Revealed by Texts from Southern Mesopotamia.* Leiden: Brill.

Leshnik, L. S. 1973. Land Use and Ecological Factors in Prehistoric Northwest India. In *South Asian Archaeology,* ed. N. Hammond, 67–84. London: Duckworth.

Mackay, E. J. H. 1931. Further Links between Ancient Sind, Sumer, and Elsewhere. *Antiquity* 5: 459–73.

Marshall, J. 1931. *Mohenjo-Daro and the Indus Civilization.* London: Cambridge University Press.

McNeill, William. 1990. The Rise of the West after Twenty-Five Years. *Journal of World History* 1: 1–21.

Moorey, P. R. S. 1994. *Ancient Mesopotamian Materials and Industries.* Oxford: Clarendon Press.

Neumann, J., and R. Sigrist. 1978. Harvest Dates in Ancient Mesopotamia as Possible Indicators of Climatic Variations. *Climate Change* 1: 239–52.

Oates, D., and J. Oates. 1976. *The Rise of Civilization*. Oxford: Elsevier.

Oppenheim, A. L. 1979. The Seafaring Merchants of Ur. In *Ancient Cities of the Indus,* ed. G. Possehl. New Delhi: Vikas.

Piggott, Stuart. 1950. *Pre-Historic India.* Harmondsworth: Penguin.

Possehl, G. 1982. The Harappan Civilization: A Contemporary Perspective. In *Harappan Civilization,* ed. G. Possehl, 15–28. New Delhi: Oxford University Press.

———. 1997. Climate and the Eclipse of the Ancient Cities of the Indus. In *Third Millennium B.C. Climate Change and Old World Collapse,* ed. H. Nuzhet Dalfes, George Kukla, and Harvey Weiss, 193–243. Heidelberg: Springer-Verlag.

Possehl, G. and M. Raval. 1989. *Harappan Civilization and the Rojdi*. New Delhi: Oxford University Press.

Raikes, Robert. 1964. The End of the Ancient Cities of the Indus. *American Anthropologist* 66(2): 284–99.

Raikes, Robert, and G. F. Dales. 1977. The Mohenjo-Daro Floods Reconsidered. *Journal of the Palaeontological Society of India* 20: 251–60.

Raikes, Robert, and R. Dyson. 1961. The Prehistoric Climate of Baluchistan and the Indus Valley. *American Anthropologist* 63(2): 265–81.

Ratnagar, Shereen. 1981. *Encounters: The Westerly Trade of the Harappan Civilization.* New Delhi: Oxford University Press.

——— 1986. An Aspect of Harappan Agricultural Production. *Studies in History* 2(2): 137–53.

———. 1991. *Enquiries into the Political Organization of the Harappan Society*. Pune: Ravish Publishers.

———. 1994. Harappan Trade in its "World" Context. *Man and Environment* 19(1-2): 115–27.

Redman, Charles. 1978. *The Rise of Civilization.* San Francisco: Freeman.

Rowton, M. B. 1967. The Woodlands of Ancient Western Asia. *Journal of Near Eastern Studies* 26: 261–77.

Sahni, M. R. 1956. Biogeological Evidence Bearing on the Decline of the Indus Valley Civilization. *Journal of the Palaeontological Society of India* 1(1): 101–107.

Schneider, J. 1977. Was There a Pre-Capitalist World-System? *Peasant Studies* 6(1): 20–29.

Shaffer, J. 1982. Harappan Culture: A Reconsideration. In *Harappan Civilization,* ed. G. Possehl, 41–50. New Delhi: Oxford University Press.

Tosi, Mauricio. 1982. The Harappan Civilization: Beyond the Indian Subcontinent. In *Harappan Civilization,* ed. G. Possehl, 365–78. New Delhi: Oxford University Press.

Wallerstein, I. 1974. *The Modern World-System.* Vol. 1. New York: Academic Press.

Wheeler, R. E. M. 1968. *The Indus Civilization.* Cambridge, UK: Cambridge University Press.

Willcox, G. 1992. Timber and Trees: Ancient Exploitation in the Middle East. In *Trees and Timber in Mesopotamia,* ed. J. N. Postgate and M. A. Powell, 1–31. *Bulletin on Sumerian Agriculture,* vol. 6. Cambridge, UK: Cambridge University.

Yoffee, Norman. 1988. The Collapse of Ancient Mesopotamian States and Civilization. In *The Collapse of Ancient States and Civilizations,* ed. N. Yoffee and G. Cowgill, 44–68. Tucson: University of Arizona Press.

6

Economic Ascent and the Global Environment: World-Systems Theory and the New Historical Materialism

Stephen G. Bunker & Paul S. Ciccantell

INTRODUCTION

The manipulation and reorganization of the relationship between nature and society is the most complex task confronting any ascendant economy. Gaining secure, inexpensive access to the huge volume of raw materials building blocks of capitalist industrial production requires economic, political, technical, and organizational innovations that restructure both existing social relationships (e.g., core-periphery relations) and the characteristics of the nature-society nexus (e.g., what raw materials are extracted where and by whom). The strategies of states and firms in ascendant economies to accomplish this task create what we term "generative sectors"; leading economic sectors that are simultaneously key centers of capital accumulation, bases for a series of linked industries, sources of technological and organizational innovations that spread to other sectors, and models for firms and for state–firm relations in other sectors. These generative sectors in raw materials and transport industries have driven economic ascent throughout the history of the capitalist world-economy.

Our analysis of economic ascent requires the reframing of world-systems theory in terms of what we call the new historical materialism. Our argument is that the distinctive feature of the capitalist world-economy is the systematic expansion of the exploitation of nature via a division of labor on an increasingly global scale. This also does not mean that this was the first time this effort had been undertaken; earlier expanding core powers and empires had sought to intensify agricultural production and to expand their raw materials supply systems. The key difference was the intensification and extension of capitalist extraction around the globe, beginning in the "long sixteenth century" and sharply increasing in scale and scope from the mid-eighteenth century onwards, restructuring social relations and the relationship between society and nature in support of capital accumulation.

This chapter will first examine the role of generative sectors in the economic ascent of Holland, Great Britain, the United States, and Japan using the method we term the new historical materialism. The striking similarities of these nations' processes of economic ascent via these generative sectors provide the basis for a nomothetic explanation of economic ascent in which the natural environment and the regularities governing material characteristics and processes play central roles. The ascent of each of these nations has restructured the capitalist world-economy by restructuring the underlying relationship with nature; in a very tangible sense, economic ascent is both a social and material process rooted in the global environment.

MATERIAL PROCESSES AND ECONOMIC ASCENT: HISTORICAL STRATEGIES

Within the capitalist world-economy, ascendant national economies require expanding access to cheap and secure sources of raw materials to sustain their challenge to established industrial economies. Lowering raw materials costs is critical to competition in international markets and is particularly important to the ascendant economy because it is also extending productive and transport infrastructure faster than the average of the established economies. Stability of supply is required for operating plants at full capacity; this is particularly important in the heavy industries in an ascendant economy because these industries involve higher than average fixed capital investments and inflexible sunk costs.

Because the states and firms of established industrial economies have often already succeeded in structuring global raw materials markets to their own advantage, the state and firms of the ascendant nation have to restructure these markets in order to compete effectively. Such restructuring is likely to collide with environmental and spatial constraints imposed by the physical characteristics of the raw materials and the location of their sources. Previously ascendant, and still dominant, economies will have organized raw materials markets in such a way as to reduce their own costs and increase their own security of supply. The established market systems are therefore likely to accommodate the organization and location of extraction, processing, and transport to the natural features and locations of natural resources and their raw material forms. The ascendant economy must therefore find new ways to accommodate to natural characteristics and to use these so as to loosen or restructure markets already built around these natural features.

Historically, ascendant economies have done this via several strategies. The first strategy is direct conquest of resource-rich peripheries, followed by wars or diplomatic actions that impede access by the established economies. The second strategy is to incorporate new technologies that effectively change established relations between economy and environment. These can include new forms or expanded scale of mining, processing, and transport. The third strategy is to induce host countries to assume a significant share of the cost of reorganizing world markets, introducing new technologies, and developing new transport routes.

These three major strategies have evolved historically to allow ascendant economies to continue their advance. The first strategy has an extremely long history, predating the emergence of the capitalist world-economy. Direct imperial conquest of resource rich peripheries and the defense of these formal and/or de facto annexations by force and/or diplomatic actions, such as Belgium's conquest of the copper-rich Congo region of Africa (Packenham 1991), have, however, become increasingly difficult and expensive to carry out and maintain.

The second strategy has been employed in a number of instances, including the adoption of James Watt's vastly improved steam engine to remove water from coal mines; Britain's relatively early industrialization based on low-cost coal was an essential element of Britain's rise as a hegemonic core power. Similarly, the rapid expansion of the domestic railroad transportation infrastructure in the United States in the mid-nineteenth century linked the United States' widely dispersed raw materials and agriculture-producing peripheries to markets and industrial centers in the East. This creation of a low-cost transport network was a central part of the United States' rapid industrialization, the key to United States ascendance in the world-economy.

The third strategy has a similarly long history in the capitalist world-economy. Raw materials producing nations have long been induced (and sometimes forced) by ascendant core powers to pay a significant share of the costs of reorganizing world markets, introducing new technologies, and developing new transport routes. Imperial core powers, for example, taxed their colonies to support armies to control indigenous populations and used corvée labor to construct infrastructure. Even in nonimperial situations, ascendant core powers have been able to induce raw materials extracting peripheries to finance the construction of railroads, for example, often justified in terms of local economic development but mainly benefiting foreign investors and raw materials consumers. Numerous examples of the employment of this strategy by Britain occurred in Latin America during the nineteenth century (Coatsworth 1981; Duncan 1932; Lewis 1983). Similarly, British and North American rubber buyers and consumers were able to induce members of the economic elite in the Brazilian Amazon to finance the expansion of the wild rubber industry in the region to supply the core's industrial plants in the late nineteenth century (Bunker 1985). This strategy dramatically reduces both the costs to and risks assumed by the ascendant core economy's firms and state in the raw materials extracting region.

Because these propositions relate to the location of the extraction, processing, and ultimate transformation of huge amounts of matter and energy, they have implications for both the global environment and a large number of specific local environments, as well as for the economic activities of human populations. Because a key component of any national raw material access strategy involves the construction of efficient transport networks on a global scale, successful strategies to restructure global raw materials markets also reorganize the global environment. Finally, these strategies may bear directly on the benefits and prejudices to human populations in natural resource exporting societies. Let us now turn to an examination of four key examples of economic ascent based on generative sectors that have restructured relations between nature and society: Holland, Great Britain, the United States, and Japan.

HOLLAND'S ECONOMIC ASCENT

Transport is the "circulatory system" of the capitalist world-economy and the process of capital accumulation; the economic ascent of Holland and later Great Britain provide particularly dramatic examples of this phenomenon. Transport industries have in many periods and nations been a focus of capital accumulation itself, and transport is in all periods the link that binds extraction, production, consumption, and waste disposal. The period commonly referred to as the mercantilist era in Europe is better characterized as the era of transport capitalism, with shipping and shipbuilding industries at the center of capital accumulation and the technological, organizational, and institutional innovations that provided the foundation for the economic ascent of Holland and later Great Britain to hegemonic positions in the capitalist world-economy (Bunker and Ciccantell 1995a, 1995b).

While trade in high value, low volume luxury goods has been the central focus of much of the analysis of this period, cursory examination of the material composition of transmaritime trade during the sixteenth and seventeenth centuries shows that the transport of bulk goods was far greater than the trade in preciosities (Wallerstein 1982; Nef 1964). While the greatest profits per voyage were clearly made in the trade in preciosities, far more boats, and thus boatbuilders, sailors, stevedores, and other linked industries were involved in the bulk trades. Linkage and spread effects from boat building and ship repair, as well as ship provisioning, were far greater in the bulk trade than in the trade of preciosities.

The location of the Netherlands as an entrepôt for exports from and imports into a vast European hinterland, in combination with the vast and diverse timber resources available in this hinterland and the early position of the Dutch as a colony of the Spanish Empire, made the Netherlands a center of shipbuilding and shipping, particularly for the movement of large volumes of bulky, low-value raw materials. The grain and wood supplies available in the Rhine and Baltic regions led to the transformation of these regions into raw materials peripheries to supply capital accumulation in the Netherlands.

The characteristics of wood shaped Holland's economic ascent in important ways. Easily transported down rivers, it is very costly to transport wood on the open seas. Bulky, rigid, and heavy, it required large ship tonnage to transport, was difficult to load and unload, and made ships both top-heavy and rigid, thus making them more likely to break up in heavy seas. Insurance and labor were both costly and difficult, especially as shipowners tended to risk only older boats in the trade. The Dutch could move timber to shipyards without an ocean voyage; even from the Baltic, the Dutch could sail along the coast, and they developed a cheap, capacious flat-bottomed boat, the Fluyt, that could move timber very cheaply. This advantage meant that it was ultimately far cheaper to build boats in Holland and sail them to Spain or Portugal than it was to ship timber overseas. The Dutch used this advantage to develop a highly sophisticated boat building industry, with cranes, winches, wind-driven sawmills, and experienced craftsmen that further increased their competitive advantage. As the Dutch expanded the control of the herring and the grain trades, as well as the reexport of Mediterranean wines and their own finished textiles, the different parts of the economy stimulated each other (Wallerstein 1982).

Wallerstein (1982) and other analysts of Dutch economic ascent have argued that the textile trade explains Dutch dominance in the Baltic trade, since rates of return were highest in textiles and because Dutch productivity in textiles retarded British development in the textile sector. Barbour (1950) shows that Dutch ships dominated traffic through the sound long before the textile trade became important, and both Barbour (1950) and Wilson (1973) place shipping and shipbuilding at the center of a complex mix of entrepôt trade, manufacture, and finance that lifted Amsterdam to economic preeminence. Wilson argued that shipping and shipbuilding constituted the major sources of linkages and multipliers as well as the critical source of raw materials to supply Dutch ascent: "There seems an incontestable case for arguing that the richest society so far in history had been the creation of sea transport" (Wilson 1973, 329). In short, the critical comparative advantages underlying Dutch ascent were its geographically provided control over river routes to the agricultural lands and forests of Poland and Germany. The shipbuilding and shipping industries based on these natural conditions became generative sectors that spread technological innovations in labor-saving machinery, organizational techniques, the adaptation of windmill technologies to wood sawing, economies of scale in protoindustrial shipyards, and the development of linked industries of finance, warehousing, and other industries like textiles that could benefit from these types of innovations.

Shipbuilding and shipping based on the competitive advantages provided by Holland's raw materials peripheries provided the foundation upon which Dutch hegemony was constructed. At the same time, these generative sectors were restructuring agricultural production systems and the use of timber in the Rhine and Baltic regions, reshaping these areas into extractive peripheries that exported raw materials to Holland.

GREAT BRITAIN'S ECONOMIC ASCENT

Rising economies attempt to foster the construction of global transport systems in patterns that reduce the costs of the raw materials they consume. The lower the value to volume ratio of raw materials, the more critical this cost reduction becomes for economic competitiveness. The sectors that pushed the development of large-capacity sailing ships with lower sailor to tonnage ratios were the timber and coal industries. National economic dominance in the preindustrial and early industrial period was closely linked to maritime trade in wooden boats, to naval security, again in wooden boats, and to wood-fueled metallurgy. In addition to depending on water for cheap transport of wood, many of the early technical advances in both wood processing and metallurgy depended on water power to drive sawmills and to power the bellows and hammers that increased fuel efficiency and labor productivity in iron smelting. Agricultural products also moved more cheaply by water, though the savings were not as important as in shipbuilding. The importance of water transport meant, though, that wood and agricultural land near watercourses were highly prized and that the various entrepreneurs who required timber extraction for metallurgy, boatbuilding, and a series of other uses competed with each other and with agriculturalists, as well as with representatives of the state, for control over and access to wood. Albion (1926) points out that naval requirements for timber competed with both corn and iron, and Ashton (1964) described what he termed the tyranny of wood and water in the development of the iron industry. This conflict over the use of internal peripheries would persist until the development of techniques for using coal to smelt iron ore.

In contrast to the Dutch intense focus on the development of cheap transport, the British specialized in the development of warships that were used to displace Holland as the economic and political center of the capitalist world-economy. Britain became increasingly dependent on the import of bulk goods in foreign ships, and the tremendous value added that transport created provided the incentives for the Navigation Acts (Davis 1973). These acts were notoriously unsuccessful at limiting the competition from the more efficient Dutch fleets. It was only through the capture of thousands of Dutch boats that the British bulk carrying trade became competitive (Davis 1973). The British maritime industry could only develop through capture because of its cost disadvantage in relation to the Dutch.

British boatbuilders built for strength and maneuverability, which required longer lines, sacrificed cargo space, and larger and more complicated rigging, and thus required more men per ton. This kind of building was useful for defense and capture, but was not particularly efficient. The British had the military edge, but not the carrying edge. In this sense, the struggle was between the locational advantages of the Dutch and the technical advantages in building and management that they accumulated through their ability to build many boats, and the bellicose strategies of the British who were compensating for their locational disadvantage with state-supported initiatives toward military prowess. Boxer (1965) attributes Dutch decline to wars, inflation, and the flight of capital into finance, but their real advantage was perhaps in location, which allowed them to develop their extraordinary entrepôt trade in heavy goods; war with Britain essentially restricted their trading advantage, and thus there was far less incentive to sustain the boatbuilding industry.

Weber (1981) and Cipolla (1965) both link the development of the metallurgical industry to the military requirements of protecting transport. The dynamic mercantile development of Britain was very much the source and the result of the timber problem there. Trade drove production and generated the income needed to stimulate it, as well as stimulating the need for armaments to protect shipping, limit rival shipping, and keep open sources of raw materials. Britain's military needs were themselves the result in large measure of continental attempts to limit British trade. The rapidly developing iron industry had been almost stagnant until the crown decided to promote domestic production of cannons and the establishment of smelters (Cipolla 1965). As that industry expanded, so did its consumption of oak, leading it into conflict with both farmers and the royal navy. The growing costs of administration, including the support of a navy necessary for the security of trade, drove the crown to look for new sources of revenue; a particularly easy one, in the short term, was the sale of trees from royal forests to the more dynamic industrial interests. Thus the burning of oak for smelters was a very hot issue during most of the seventeenth century. The success of the mercantile economy, and its demand for inputs, stimulated other economies that required the same raw materials. Access to foreign timber thus became critical for multiple purposes, especially in dramatic surges of demand such as that occasioned by the London fire, which Albion (1926) characterizes as warming Finland's economy. Britain suffered from the limitations of her own supplies and the distance to other sources.

The adoption of James Watt's vastly improved steam engine to remove water from coal mines in Great Britain during the last 20 years of the eighteenth century began a shift away from wooden shipbuilding and toward the development of internal canal and railroad transport and iron industries as generative sectors. Watt's steam engine made vast reserves of deeply buried coal that had previously been unextractable both technologically and economically suddenly available on a large scale at low cost to power Britain's industrial revolution. A massive canal building effort to link internal coal fields to industrial and population centers became a major focus of capital accumulation in Britain (Mathias 1969, 134–35; Rosenberg and Birdzell 1986, 150–51).

Within a few decades, the steam engine was adapted for railroad transport, simultaneously freeing Britain from increasingly expensive, complex efforts to build canals to supplement natural watercourses and creating a massive synergy between railroad transport and the iron and later steel industries during the nineteenth century. British ironmasters discovered the sulfur-reducing chemistry required to smelt iron with coal and progressively reduced coal charges per unit of production, as well as developing the Bessemer converter in 1856 that made mass production of steel possible. Further, the British development of the Siemens-Martin open hearth furnace that increased productivity and the widening of the range of ores from which steel could be made via the development of the Gilchrist-Thomas basic process (Isard 1948; Hobsbawm 1968) were also examples of the role of the coal, iron, and linked transport industries as generative sectors driving British economic ascent and hegemony. Moreover, these developments further tied Britain's internal raw materials peripheries to the process of capital accumulation. In the mid-nineteenth century, the steam engine was also applied to water transport, rejuvenating Britain's shipbuilding and shipping industries on the basis of steamships that linked the distant parts of the British empire; shipbuilding was a massive consumer of raw materials that were often transported on steamships themselves (Mathias 1969; Rosenberg and Birdzell 1986). The combined impacts of railroads, steamships, and the raw materials industries on which they depended were to revolutionize industry and finance in Britain (Hobsbawm 1968), becoming generative sectors that drove Britain's economic ascent.

Britain's relatively early industrialization based on low-cost coal was an essential element of Britain's rise as a hegemonic core power. This is the essence of the role of generative sectors in economic ascent; what might be termed "virtuous cycles" of linkages between raw materials and transport industries drove capital accumulation in Britain during its phases of economic ascent, based on incorporating first internal and later external peripheries, and during its period of hegemony.

From the perspective of the raw materials periphery in which inland transport systems and ports to export raw materials to a core power such as Great Britain are located, these generative sectors and the transport networks thus developed have very different impacts. Innis (1956) demonstrates the relationships between core demand for raw materials, the transport infrastructure required to satisfy that demand, the financial instruments and agencies required to finance this infrastructure, and the forms of governance necessary to assure the payment of debts incurred to build this transport infrastructure. Innis links Canada's Articles of Confederation directly to the financing of railroads and rebellions against the state to regional competition for transport. The notorious, and eventually abandoned, demand by British capitalists that Latin American nations guarantee a minimum rate-of-profit for railroads built to move raw material and agricultural products to exporting ports made similar demands on these nation-states.

More generally, the nation-state, its control over its own territory, and its taxation and borrowing powers appear in many instances as a hegemonically imposed device to assure the huge sunk capital needed to create the globally built environment that Britain needed to channel adequate supplies of matter and energy to its rapidly growing industries (Adams 1982). Nationhood as a desired goal of Spanish colonies in the Americas was to Britain a hegemonically useful ideology. Canada became more autonomous from Britain precisely to allow it to assume the costs and guarantee the loans required to dredge canals, build locks, and construct railroads to allow the large scale export of raw materials to Britain. The construction of railroads in India and Afghanistan required and then molded changes in local states and in the relations between them. In all of these cases, local social relationships were restructured to permit the extraction of raw materials to support capital accumulation in Great Britain.

U.S. ECONOMIC ASCENT

Britain's growing reliance on the agricultural and industrial development of the American colonies, particularly the development of the New England shipbuilding and shipping industries, laid the foundation for the economic ascent of the United States. Abundant U.S. timber supplies, numerous river networks to transport timber to the coast, the transport cost advantages of processing timber into ships at the rivers' mouths rather than shipping to English shipyards, and the United States' status as a British colony gave U.S. shipbuilding and shipping industries a tremendous competitive advantage in the world-economy. These industries were generative sectors in the eighteenth and first half of the nineteenth centuries, transporting bulk and luxury goods over long distances to Europe, China, and other parts of the world.

An important difference between New England's and the earlier case of Holland's ascents based on shipbuilding was that Holland's hinterland had been relatively densely settled for centuries, and its societies were constituted into political units capable of significant defense and aggression. Thus, bellicose expansion of territory was impossible and wars, particularly land wars, extremely costly. The United States enjoyed a hinterland whose earlier occupants had been severely dislocated and were progressively diminished in number and political strength. Thus, it was ultimately possible in the United States to combine raw materials sources and industrial centers within the same sovereign unit. This pattern would later be replicated via canal and then railroad building, incorporating raw materials rich regions as internal United States peripheries.

Perhaps even more important in the early phase of United States economic ascent was that the production of the various bulk goods exported from the United States did not require huge capital outlays. In the Europe–Asia trade, for example, the cargo itself might be worth ten times the value of the boat itself. Such trade was only accessible to highly capitalized merchants. Returns on exports of wood and cotton cargoes or on shipments of ice and granite from the United States might return far less but would pay a return on the shipping itself and were therefore accessible to the smaller capitals required to build and man a ship. United States ships were for a long time smaller than European ships, especially in the Far Eastern trade, again resulting in the reduction of the total capital risked. United States trade to China started with sea otters from the Northwest; huge returns from this trade were derived from transport rather than from the cargo itself. The lack of high-capital barriers to entry and the large returns available to shipping in many export trades from the United States meant that the transport business could be more decentralized, both in terms of location and in terms of ownership.

The other peculiar advantage of topography for the United States was the number of rivers flowing out of the Appalachians which could be dammed to produce power. These rivers powered textile and shoe mills, as well as sawmills. Shipbuilding skills and labor and merchant capital were drawn to New England by cheap timber and trade opportunities. Shipbuilding requirements also included a variety of other inputs in addition to timber, including nails, block and tackle, and sails, demand for which led to the creation of linkages to ironworking, sail making, and other industries which supplied these essential inputs to local shipyards. This particularly favorable coincidence of natural conditions with the leading economic sectors in the world-economy of the period (shipping and shipbuilding) and the changing political context of the eighteenth and nineteenth centuries provided the foundation for the early economic ascent of the United States. However, United States shipbuilding and shipping industries were made uncompetitive by the large-scale introduction of British steamships that restored Britain's control of ocean transport.

The rapid expansion of a domestic transportation infrastructure in the United States in the mid-nineteenth century based on the newly developed technology of railroads served to link the United States' widely dispersed raw materials and agriculture-producing peripheries to markets and industrial centers in the East (Stover 1961; Chandler 1965; Douglas 1992). Waterways and later railroads led to the incorporation of a wide range of domestic raw materials peripheries, including agricultural products and later coal in Appalachia (Dunaway 1996), copper in Michigan (Leitner 1998) and later Montana and the Southwest, and iron ore in Michigan and Minnesota, among many others. The railroad network was also extended to incorporate Mexican and Canadian raw materials peripheries that supplied a diverse set of raw materials to the United States (Ciccantell 1995). This creation of a low-cost transport network was a central part of the United States' rapid industrialization, the key to U.S. ascendance in the world-economy.

This incorporation of internal, nearby, and increasingly distant raw materials peripheries to supply United States economic ascent also transformed social relationships in these peripheries, restructuring them to provide labor for extraction, while areas previously populated by indigenous groups or used for farming and ranching were transformed into sites of extraction. United States involvement in anticolonial movements and other interventions to create and maintain raw materials peripheries and transport systems have provided similar hegemonic benefits; like the Suez Canal before it, the construction of the Panama Canal and the aborted negotiations for a canal in Nicaragua involved the creation or subordination of nation-states. In recent years, the Canada–U.S. Free Trade Agreement and the North American Free Trade Agreement represent attempts to reconstruct U.S. hegemony by restructuring U.S. raw materials supply networks via renewed, lower-cost access to raw materials in Canada and Mexico (Ciccantell 1997).

JAPAN'S ECONOMIC ASCENT

Japan's ascendance from the periphery to the core of the capitalist world-economy began during the Meiji period in the last third of the nineteenth century. Confronted by powerful economic and geopolitical rivals in the Pacific region, including the United States, Russia, China, and the European colonial powers (see McDougall [1993] for a discussion of the history of this geopolitical rivalry), industrial development became the basis of Japanese economic and military strategy (see, e.g., Nafziger 1995). Japanese efforts to industrialize and build a strong military paid early dividends in the form of victories in the Sino-Japanese War of 1894 and the Russo-Japanese War at the beginning of the twentieth century. The Sino-Japanese War also gained for Japan its first formal and informal colonies of the modern era, as well as indemnification that helped finance the expansion of the iron and steel industries in Japan (So and Chiu 1995, 89–90).

Much of Japan's success was, however, due to its ability to export light industrial products such as silk and to use the proceeds to import both ships and steel plates for building military and trading ships (Chida and Davies 1990). Producing consumer goods for domestic consumption and export, often by importing technological advances and then improving and adapting them for new uses, has remained an important engine of the Japanese economy. However, even these industries are critically dependent on the availability of raw materials used in their production.

Efforts to deepen industrialization in Japan were undertaken during the first third of the twentieth century, most notably through expanding the steel, copper, and shipbuilding industries and through the creation of a domestic aluminum industry. This industrialization drive rapidly depleted Japan's limited coal, iron ore, and copper reserves. In order to support the rapid industrialization drive in the years between the First and Second World Wars, the Japanese state and firms sought to gain access to raw materials that were being rapidly depleted in Japan via the first strategy for continuing its ascendance in the world-economy, direct imperial conquest of neighboring resource-rich areas of China, East Asia, and Southeast Asia (So and Chiu 1995). However, this raw materials access strategy brought Japan into direct military conflict with the United States, Great Britain, the Soviet Union, and China. As historian Marshall (1995, x) has argued, "the United States' war with Japan from 1941 to 1945 was primarily a battle for control of Southeast Asia's immense mineral and vegetable wealth." The results of this conflict were the defeat of Japan in World War II, the dismemberment of Japan's empire, and severe economic and political crises in Japan in the war's aftermath. Japan's defeat in World War II foreclosed this ascendance and development strategy.

The Japanese steel mills, with the assistance of the Japanese state, devised a model to guarantee long-term secure access to metallurgical coal and iron ore from Australia, the closest nearby politically available source of raw materials for what Japanese and American military planners hoped would prove to be a generative sector for Japan's renewed economic ascent. The Japanese steel mills utilized a new model of long-term contracts, at first forced upon them by Australia and the United States, rather than using the wholly-owned foreign direct investment model utilized by U.S. and European steel firms to gain access to foreign raw materials sources. This new model accommodated the resource nationalism of host nations such as Australia, while in the process restructuring worldwide flows of metallurgical coal from mainly domestic movement from captive mines to their steel mill owners to transoceanic trade flows governed by long-term contracts, fundamentally altering the nature and composition of the world metallurgical coal industry. Metallurgical coal was extracted by Australian and transnational firms, which assumed the capital cost and risks of opening up previously unexploited coal deposits, deposits which had not even been explored for earlier because of the tremendous distances between these deposits and potential markets. The coal was transported by Australian state-owned railroads to typically state-owned ports, transferring the capital and risk burden to the raw materials periphery local and national governments. At the state-owned ports, the coal was loaded on Japanese ships for the trip to Japan.

The natural availability of metallurgical coal, United States-led diplomatic efforts, and the development of long term contracts are only part of the story; coal had been acquired, but how could millions of tons be moved to the new coastal steelworks in Japan at low enough cost? Two other natural characteristics of the Australian continent, the location of these coal deposits less than 300 kilometers from the eastern coast of Australia and the characteristics of the Australian coast that permitted the construction of large-scale ports for ore carriers made it possible for the Japanese to promote a fundamental restructuring of space and nature via transport technology. The costs of the rail transport infrastructure were borne by the state government (Frost 1984, 49–53). The port facilities were typically built and operated by the mining companies themselves (Frost 1984; IEA 1992, 109; Tex Report 1994b, 552–55), although some ports were later built by state governments (IEA 1992, 109).

This transport pattern allowed Japanese steel mills and shipping firms to take advantage of the tremendous economies of scale available in bulk shipping to dramatically reduce production costs of steel in Japan by capturing all of these benefits for themselves. The key elements of transport as a raw materials access strategy have included research and development on the construction of larger petroleum tankers and bulk carriers and the construction of large shipyards capable of building such large ships. These large ships are owned and operated by Japanese shipping firms associated with the major industrial groups; these Japanese industrial groups control ocean shipping of raw materials on an FOB raw materials exporting port basis so that any reductions in transport costs caused by technological improvements or changes in world shipping market conditions are captured by Japanese importers. The construction of large-scale port and railroad infrastructures in raw materials exporting regions paid for by extractive region governments and/or raw materials transnational corporations is based on long-term contracts for raw materials supply with Japanese importing firms to allow the efficient use of these large ships.

Additionally, the Japanese government provides subsidies for the construction of maritime industrial areas in Japanese ports which eliminate the need for internal transshipment in Japan of raw materials imports (Bunker and Ciccantell 1995a). Japan's coastline was ideally suited for this form of linkage and transport-based development (Kosai and Ogino 1984, 60–61).

Capturing economies of scale in transport requires the construction of massive port systems, capable not only of accommodating large boats, but also of loading them and unloading them quickly enough to prevent incurring the huge costs of tying up the capital-intensive ships too long in harbor. The costs of building such ports have enhanced a feature of all constructed transport systems, that is, that to the extent that exporting and importing systems must be physically compatible to take advantage of cost-saving technologies, importers can tie exporters to their markets by fomenting mutually compatible port systems at both ends of the voyage. These investments in large-scale ports physically and economically tie raw materials exporters to only a very small number of potential customers, almost all of them located in Japan and Western Europe (Sullivan 1981), because the high capital investment in large-scale ports and mines can only be repaid by a high rate of capacity utilization. A high rate of capacity utilization is dependent on the use of large-scale ships in these large-scale ports. This natural and social restructuring has converted Australian, Brazilian, Canadian, and other raw materials rich regions into raw materials peripheries supplying capital accumulation in Japan.

These economies of scale in raw materials extraction and transport are tightly linked to economies of scale in steel production itself. Abegglen and Stalk (1985) argue that these three types of economies of scale, including the construction of new integrated steel mills in Japan from the 1950s to the 1970s that when built were the largest or almost the largest in the world, gave Japan a tremendous competitive advantage in the world steel industry. Because Japan lacked domestic supplies of metallurgical coal and iron ore, Japanese steel firms were able to search out and help develop the lowest-cost suppliers in the world which had access to large-scale ocean shipping potential, resulting in significant raw materials cost advantages for Japanese steel firms (Abegglen and Stalk 1985, 73–78). As American and Japanese development planners foresaw in the late 1940s, the steel industry has become the linchpin of a number of linked industries which have complemented one another in a "virtuous cycle" of economic development based on generative sectors in shipbuilding and steel, transforming Japan into the world's second largest economy and the United States' most formidable economic competitor. This pattern of metallurgical coal supply relationships has also been replicated in a number of other raw materials peripheries around the world.

While this pattern was well suited to Japanese needs and initially allowed Japan to resume trade with Australia despite Australian antipathy toward Japan, this transfer of capital costs and risks to exporting firms and nations has often proven to be quite deleterious to these firms' and nations' interests in the long term (Koerner 1993), even though the original idea for these long-term contract arrangements came from the Australians (Priest 1993, 20–25). Similarly, huge investments in railroad and port facilities have generated limited returns for extractive peripheries.

In summary, the Japanese steel mills and the Japanese government, with initial support by the existing hegemon, the United States, have succeeded in restructuring the world metallurgical coal industry and other raw materials industries to support Japanese industrialization. This restructuring was a fundamental material and economic pillar of Japan's rise as an industrial power and challenger to U.S. economic hegemony, based on these transformations of social relations and society-nature relations in these raw materials peripheries supplying Japanese economic ascent.

CONCLUSION: WORLD-SYSTEMS THEORY, THE GLOBAL ENVIRONMENT, AND THE FUTURE OF THE CAPITALIST WORLD-ECONOMY

A great deal of attention has been devoted (at least since the Club of Rome report of the 1970s) to the proposition that the local and increasingly global environmental destruction and natural resource depletion underlying the capitalist world-economy are leading to a systemic collapse. The process of capital accumulation based on the exploitation of nature has been exceeding the ability of nature to replenish natural products and absorb waste for at least the last several decades. Sooner or later, this relationship between the capitalist world-economy and nature will destroy the natural bases on which this social system depends. In a matter of years or decades, this line of reasoning argues, the capitalist world-economy will either be transformed into a more ecologically sensitive and humane system or the earth's ecosystem will collapse.

The analysis of economic ascent and recasting of world-systems theory in terms of the role of material processes lends credence to this line of argument by highlighting the systematic, expanding dependence of the capitalist world-economy on the exploitation of nature. As the historical examples discussed above amply demonstrate, economic ascent and, more broadly, the process of capital accumulation have entailed a continually expanding process of the depredation of nature. However, this seemingly obvious conclusion drastically underestimates the powerful incentives for innovation and adaptation in the capitalist world-economy and the fungibility of the relationship between society and nature.

What does this mean? To return to an earlier example, 50 years ago it was completely uneconomic to move coal more than a few hundred miles to generate electricity or to produce steel; today, tens of millions of tons of coal are moved each year from remote, coal-rich peripheries to fuel core industries. A seemingly incontrovertible characteristic of the relationship between society and nature, that centers of coal consumption must be located near the naturally determined locations of coal deposits because of the huge costs of moving this bulky material, had been annihilated by innovations in long-distance ocean transport and land reclamation for industrial use in ocean-accessible coastal areas as part of Japanese strategies for economic ascent.

Another trenchant example of innovation and fungibility is the current interest in superconducting materials. While usually discussed in terms of high-technology applications, perhaps the most important large-scale application of superconducting materials would be to permit the long-distance transport of electricity. Innovations over the last 30 years have sharply increased the distance that electricity can be transported. The development of commercial superconducting transmission lines would allow dozens of hydroelectric dams in the Amazon or nuclear power plants in lightly populated Arctic regions to supply electricity throughout North America and coal-fired or nuclear power plants in Siberia to supply Japan and Europe, relocating the environmental costs and consequences of core capital accumulation to these remote peripheries.

These potential innovations and fungibility represent core powers' and firms' efforts to maintain or enhance their positions in the capitalist world-economy by reshaping the relationship between society and nature, just as does the emerging traffic in toxic waste exports to Africa to reduce the costs of disposing of core wastes by transferring the burden to the periphery (Frey 1998). Free trade agreements like NAFTA and the WTO that tighten the links of incorporation between raw materials and low-cost labor peripheries and core powers are another example of restructuring social relations and reshaping the relationship with nature in support of core capital accumulation.

High-technology industries, service sectors, and financial machinations are obviously central components of the capitalist world-economy; what is less obvious (in what is often mistakenly labeled as a "postindustrial" or "dematerializing" or "information economy") are the material foundations on which this world-system is built. The recasting of world-systems theory to highlight its material and environmental bases via the new historical materialism provides a framework for understanding the distinctiveness of the capitalist world-economy in comparison with earlier periods, the material foundations of core economic ascent and its obverse, the incorporation of peripheral regions and peoples in support of core capital accumulation, and the possible future of the capitalist world-economy as we near the twenty-first century.

REFERENCES

Abegglen, James, and George Stalk. 1985. *Kaisha, The Japanese Corporation.* New York: Basic Books.

Adams, R. 1982. *Paradoxical Harvest: Energy and Explanation in British History, 1870–1914.* Cambridge, UK: Cambridge University Press.

Albion, R. 1926. *Forests and Sea Power: The Timber Problem of the Royal Navy, 1652–1862.* Cambridge, MA: Harvard University Press.

Ashton, Thomas Southcliffe. 1964. *The Industrial Revolution, 1760–1830.* New York: Oxford University Press.

Barbour, V. 1950. *Capitalism in Amsterdam in the Seventeenth Century.* Baltimore: Johns Hopkins University Press.

Boxer, C. 1965. *The Dutch Seaborne Empire: 1600–1800.* New York: Knopf.

Bunker, Stephen G. 1985. *Underdeveloping the Amazon.* Chicago: University of Chicago Press.

Bunker, Stephen G., and Paul Ciccantell. 1995a. A Rising Hegemon and Raw Materials Access: Japan in the Post-World War II Era. *Journal of World-Systems Research* 1(10): 1–31.

Bunker, Stephen G., and Paul Ciccantell. 1995b. Restructuring Space, Time, and Competitive Advantage in the World-Economy: Japan and Raw Materials Transport After World War II. In *A New World Order? Global Transformations in the Late Twentieth Century,* ed. D. Smith and J. Borocz. Westport, CT: Greenwood Press.

Chandler, Alfred, ed. 1965. *The Railroads: The Nation's First Big Business.* New York: Harcourt, Brace and World.

Chida, Tomohei, and Peter Davies. 1990. *The Japanese Shipping and Shipbuilding Industries.* London: Athlone Press.

Ciccantell, P. 1995. Integrating the NAFTA Market: Raw Materials and Transport Industries. Paper presented at the meeting of the Latin American Studies Association, Washington, DC, September 28–30.

Ciccantell, P. 1997. NAFTA and the Reconstruction of U.S. Hegemony: The Raw Materials Foundations of Economic Competitiveness. Presented at the Annual Meeting of the International Studies Association, Toronto, Canada, March.

Cipolla, C. 1965. *Guns and Sails in the Early Phases of European Expansion, 1400–1700.* New York: Pantheon Books.

Coatsworth, John. 1981. *Growth Against Development: The Economic Impact of Railroads in Porfirian Mexico.* DeKalb: Northern Illinois University Press.

Davis, Ralph 1973. *The Rise of the Atlantic Economies.* London: Weidenfeld & Nicolson.

Douglas, George. 1992. *All Aboard! The Railroad in American Life.* New York: Paragon House.

Dunaway, Wilma. 1996. *The First American Frontier: Transition to Capitalism in Southern Appalachia, 1700–1860.* Chapel Hill: University of North Carolina Press.

Duncan, J. 1932. *Public and Private Operation of Railways in Brazil.* New York: Columbia University Press.

Frey, Scott. 1998. The Hazardous Waste Stream in the World-System. In *Space and Transport in the World-System,* ed. Paul S. Ciccantell and Stephen G. Bunker, 84–103. Westport, CT: Greenwood Press.

Frost, D. 1984. The Revitalisation of Queensland Railways Through Export Coal Shipments. *Journal of Transport History* 5(2): 47–56.

Hobsbawm, E. 1968. *The Age of Industry*. New York: Scribner's.

Innis, H. 1956. *Essays in Canadian Economic History*. Toronto: University of Toronto Press.

International Energy Agency (IEA). 1992. *Coal Information*. Paris: International Energy Agency.

Isard, W. 1948. Some Locational Factors in the Iron and Steel Industry since the Early Nineteenth Century. *Journal of Political Economy* 65(3): 203–17.

Koerner, Richard. 1993. The Behaviour of Pacific Metallurgical Coal Markets: The Impact of Japan's Acquisition Strategy on Market Price. *Resources Policy* (March): 66–79.

Kosai, Yutaka, and Yoshitaro Ogino. 1984. *The Contemporary Japanese Economy*. Armonk, NY: M.E. Sharpe.

Leitner, Jonathan. 1998. Raw Materials Transport and Regional Underdevelopment: Upper Michigan's Copper Country. In *Space and Transport in the World-System,* ed. Paul S. Ciccantell and Stephen G. Bunker, 125–51. Westport, CT: Greenwood Press.

Lewis, Colin. 1983. *British Railways in Argentina 1857–1914*. London: Athlone Press.

Marshall, Jonathan. 1995. *To Have and Have Not: Southeast Asian Raw Materials and the Origins of the Pacific War*. Berkeley: University of California Press.

Mathias, Peter. 1969. *The First Industrial Nation: An Economic History of Britain 1700–1914*. New York: Scribner's.

McDougall, Walter. 1993. *Let the Sea Make a Noise: Four Hundred Years of Cataclysm, Conquest, War and Folly in the North Pacific*. New York: Avon Books.

Nafziger, E. Wayne. 1995. *Learning from the Japanese: Japan's Pre-War Development and the Third World*. Armonk, NY: M.E. Sharpe.

Nef, J. 1964. *The Conquest of the Material World*. Chicago: University of Chicago Press.

Packenham, Thomas. 1991. *The Scramble for Africa*. New York: Random House.

Priest, R. Tyler. 1993. Coal: Australia 1946–1960. University of Wisconsin-Madison. Unpublished manuscript.

Rosenberg, Nathan, and L.E. Birdzell. 1986. *How the West Grew Rich: The Economic Transformation of the Industrial World*. New York: Basic Books.

So, Alvin, and Stephen Chiu. 1995. *East Asia and the World Economy*. Thousand Oaks, CA: Sage Publications.

Stover, John. 1961. *American Railroads*. Chicago: University of Chicago Press.

Sullivan, A. 1981. Foreign Coal Ports Expand Capacity. *Coal Age* (May): 110–15.

Tex Report. 1994a. *Iron Ore Manual 1993-94*. Tokyo: Tex Report Company.

Tex Report. 1994b. *1994 Coal Manual*. Tokyo: Tex Report Company.

Wallerstein, I. 1982. Dutch Hegemony in the Seventeenth-Century World Economy. In *Dutch Capitalism and World Capitalism,* ed. Maurice Aymard, 93–146. Cambridge, UK: Cambridge University Press.

Weber, Max. 1981. *General Economic History*. New York: Free Press.

Wilson, Charles. 1973. Transport as a Factor in the History of Economic Development. *Journal of European Economic History* 2(2): 320–37.

7

The Development of the Risk Economy in the Circumpolar North

Ilmo Massa

INTRODUCTION[1]

The circumpolar zone is one of the last relatively untouched wildernesses in the world. By the circumpolar zone I mean the area north of the 60th parallel, comprising nearly 19 million square kilometers. In this huge area only about 15 million people live, most of them in the Nordic countries whose weather conditions are relatively good because of the warm Gulf Stream (Hustich 1972, 185). In this area there are both rather isolated and almost uninhabited "passive peripheries" and "active peripheries" quite firmly integrated in the southern economic centers (Hamelin 1978, 65–66). Politically, the northern zone is divided among the Nordic countries, Canada, the United States, and Russia.

The great climatic variations are one of the most important northern features influencing the environmental history of the circumpolar zone. Climate conditions have not only been background factors of cultural development, but active factors with great historic significance. The so-called climatic hazard coefficient increases when going northwards. In the economy of the northern native peoples and settlers, great variations between "good" and "bad" summers were a major risk. In good years there was plenty of game and plant products; in bad years starvation was a dismal fact of life. The traditional diversified self-sufficiency of the northern native people and early settlers was one response to these heavily fluctuating climatic conditions.

One important result of the permanent coldness is permafrost, which has a direct effect on the utilization and regeneration of nature and its resources. Permafrost covers most of Greenland and Alaska, half of Russia and Canada, all the year around, but, thanks again to the Gulf Stream, only small areas in the Scandinavian countries. In Siberia, permafrost at its thickest reaches a depth of 1500 meters. During summer the carrying capacity of the ground is weakened dramatically by the melting permafrost, making it difficult to build even light structures. For example, pipelines, water pipes, and sewage systems have to be insulated and lifted onto supporting structures, so that their construction is very expensive.

For hundreds of years, the future of the northern economy and its environmental quality has not been decided locally but at the centers of world-economy. The permanent European expansion began in the fifteenth century when boats coming from Europe (England, the Netherlands, France, and Spain) started fishing and whaling, first in the northern Atlantic and later in the Arctic Ocean. Soon this development was associated with the fur trade with the native people. Industrialization penetrated into the circumpolar north as ecological colonialism became oriented toward the export of unprocessed raw materials, notably timber and minerals, to the centers of civilization.

Interdisciplinary Environmental History

Environmental history is defined as the study of an interaction between the human and nonhuman components of the natural world. It involves analysis on three levels, which are not conflicting but rather complementary approaches.

First is ecological history (Donald Worster's natural history), or environments of the past, which includes the history of climate, geology, plants, animals, insects, and microbes, rates of reproduction of organisms, and reproductive success and failure. Culture often remains on the margin.

Second is the history of environmental ideas, embracing the environmental history of ideologies, perceptions, values, religions, myths, philosophy, and science. This approach sees nature as a historically contingent and complex term. Forest resources, for example, are not just materials extracted from the environment through technologies but are cultural constructs, including what people think about forests.

Third comes interdisciplinary environmental history. Specifically, it includes efforts to create altogether new ecological interpretations of history. Environmental problems are embedded in a social, political, and cultural context and challenge fundamentally the existing frontiers of specialized disciplines.

The whole western economic process with its serious ecological problems can be described by the term "risk economy." Risk economy means that in industrial societies, growing risks seem to slip away from security and control mechanisms in a world-system whose development is going more and more out of control. The term "risk economy" is an analogy taken from sociological discussions about risk society originating from Ulrich Beck's (1986) book, *Risikogesellschaft* (Risk Society).[2]

Traditional economic thinking is based on the idea that economic development is a progressive phenomenon where risks, if there are any, can be managed, mainly by technocratic regulations. Until now, risk economy was only a metaphor and has not been systematically developed as an instrument of sociological or environmental–historical research. An important engine working behind the development of the risk economy is the economy of extraction described by Stephen Bunker ([1985] 1988, 58–76). By extractive economies he means export economies (or staple economies à la Harold Innis), based primarily on the extraction of value from nature rather than on the creation of value by labor, which can be called productive economies. Bunker argues that traditional economic models of industrial production neglect the extractive origins of the materials which industrial production processes transform.

Figure 7.1 offers a schematic and simplified model of the economic mechanisms behind what can be called "global risk civilization." Risk economy is seen as the most important factor drawing technological and economic development towards an extractive economy. The extractive economy has spread from industrial centers to all parts of the world, which can be called ecological or resource-based colonialism.

Figure 7.1 Global Risk Civilization

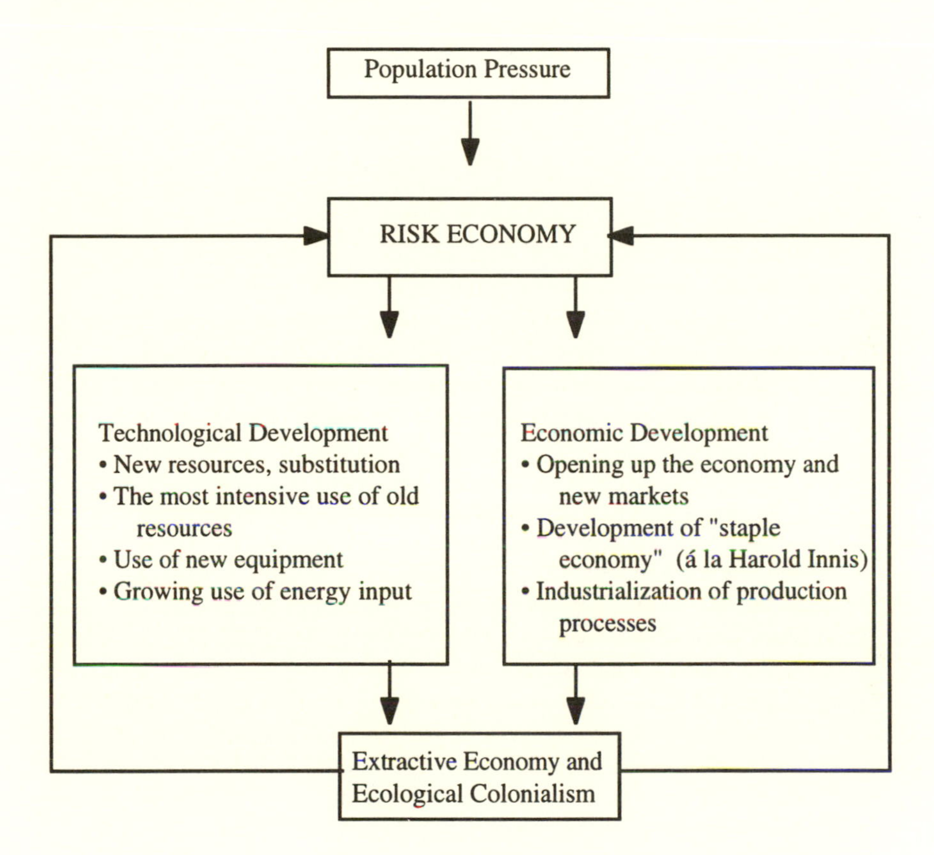

ECOLOGICAL COLONIALISM IN THE CIRCUMPOLAR NORTH

The Beginnings of Europeanization

In deeply rooted beliefs European civilization is still considered an engine that is driving the barbarian world toward civilization. The most recent version of this superiority complex is now hidden behind economic materialism. According to the ideology of economic materialism, industrialization is an essential precondition of success and the satisfaction of needs. For example, industrially produced goods are always *a priori* better than their handmade counterparts.

There has always existed on the margin of science the contradictory view that the Europeanization of the world has damaged seriously other cultures and natural settings. The modern wave of environmental history has again placed this discussion at the center. Ten years ago, Alfred Crosby (1987) published his book, *Ecological Imperialism,* which traces the biological triumph of Europe from the early Middle Ages to the present. Clive Ponting's *Green History of the World* (1992) is also an interesting attempt to write a synthesis of world history where nature is not a humble servant but a new hero.

Europeanization can be defined as the direct migrations of people from Europe to other continents and the diffusion of European culture all over the world. From the fourteenth century onwards, the Eurocentric trade system bound the world as a global economy. Europeanization laid the foundation for the first real world-economy. This huge economic mill used resources from all corners of the world forcing, at the same time, "unhistoric people" (Wolf 1982) into its jaws. Grain was taken from Eastern Europe, metals and fish from South America. Fish and whale oil were taken from the waters of North America, especially from the Grand Banks of Newfoundland. The boreal forest belt produced furs, sawtimber, and metals. In the tropics of North America and India, Europeans created sugarcane, cotton, and rice plantations. The production of spices was traditionally left to Asia.

Ponting has summed up the environmental history of a Europeanized world-system as "a destructive flood sweeping over the world." Europeans set the colonized areas to producing tea, coffee, cotton, rubber, and furs, paying practically no attention to the social and ecological costs involved. Many native peoples were decimated by the infectious diseases that came with European traders, often without even having seen the new conquerors.

Following the ill-fated Viking settlements in the early Middle Ages, more permanent European colonization of the circumpolar zone occurred in the fifteenth century when England, Holland, Spain, and Portugal started sailing in the northern waters for fur trading, whaling, and cod fishing. In the middle of the sixteenth century, English and Dutch sailors were searching for a sea route—later this was called the Northeast Passage—around Eurasia to China and India. The effort was stopped by the ice masses of the Kara Sea. The sailors founded Spitzbergen and established a fur-trading post on the coast of the White Sea. The first navigation of the Northeast Passage was made in 1878–1879 by Adolf Erik Nordenskiöld.

The early history of the so-called Northwest Passage, the sea route from the Atlantic Ocean to the Pacific, is connected with the repeated efforts to find a sea route to India. Martin Frobisher made the first attempt to find the route when mapping Baffin Bay. The final breakthrough was made by Roald Amundsen (1872–1928) who sailed with a steam-powered ship through the Canadian arctic archipelago in 1903–1906.

The North Pole remained as a last unconquered spot in the northern zone until 1909 when American explorer Robert Peary reached it. Nowadays polar areas are known better than ever thanks to increasing research efforts and the use of recent technologies such as planes, icebreakers, nuclear submarines, and satellites.

Fur Trade

In the circumpolar zone fur was the first resource to attract Europeans. The markets for furs were in Europe where, for example, beaver hats were in fashion from the end of the sixteenth century to the beginning of the nineteenth century (Braudel 1984).

In the sixteenth century, under the reign of Ivan the Terrible, Russia expanded very rapidly from its European core areas toward the Pacific Ocean, which was reached in the 1640s. At that time, furs together with grain were by far the most important Russian exports. Strong measures were taken to ensure that furs flowed continuously from Moscow to world markets (Attman 1973). For example, the Cossacks were hired to safeguard fur taxing and trading against possibly hostile native peoples. In 1581, Russians crossed the Ural Mountains and in the following 60 years conquered Siberia from the Ural Mountains to the Pacific Ocean. Kamchatka was reached in 1707, and 30 years later Russians established the first colony in Alaska.

The areas colonized by the Russians were "owned" by fifteen subarctic and arctic peoples, who lived by hunting, fishing, and reindeer herding. However, the colonization was usually peaceful because it was useless to battle with bows and arrows against Russians with firearms. As their numbers rapidly grew, the Russians began to seriously disturb the native way of life. By 1700, the Russians were already a majority in Siberia (Armstrong 1958, 114).

The northern expansion of the Russians has often been seen as a proof of the typical dynamics of Russian culture. However, one can find a more realistic historical explanation for the Russian expansion. Fur traders had constantly to move to new hunting areas because the most accessible animal populations were hunted with firearms and other effective hunting techniques. In effect, the fur trade was clearly an extractive business. Probably no species were hunted out, but there is much information on the diminishing productivity of hunting (Armstrong 1965, 61).

At least since the middle of the nineteenth century, information that the hunting of whales and sea mammals was becoming unprofitable had also been coming from Alaska. This is the most important reason why the Russians decided to sell the whole of Alaska to the United States for only a little over seven million dollars (Morehouse 1984). However, the same kind of ecological colonialism continued in Alaska after it was acquired by the United States (Armstrong, Rogers, and Rowley 1978, 129–32).

In Canada, the ecological history of the fur trade has been documented perhaps better than anywhere else in the circumpolar zone. At the beginning of the seventeenth century, the fur trade was only a small sideline to the fishing off the Grand Banks in Newfoundland. Yet, by the middle of the nineteenth century, the whole of Canada was covered by fur-trading networks laying the foundation for the Canadian federation (Innis [1930] 1956, 9).

The three main fur-trading routes were the Saint Lawrence River, the Hudson River, and Hudson Bay. Most important was the Saint Lawrence, originally controlled by France. This huge river system, along with tributaries and the Great Lakes, opened up the internal parts of Western Canada for fur trading. Another very important route was the Hudson River, which was first controlled by Holland and then England. The third route was Hudson Bay, discovered in 1668. Around the Hudson Bay fur trade, a new company, Hudson's Bay Company, was established. It was to become one of the most important institutions in the history of Canada (Innis [1930] 1956, 12–13).

Innis ([1930] 1956, 42) has described the collapse of the beaver population in Canada as "a catastrophe sweeping over fur-trading areas of North America." Beaver was easy to kill and new hunting techniques—firearms, steel traps, and the use of castoreum as a decoy—made beaver hunting more efficient. In the seventeenth century, the beaver was wiped out from the valley of the Saint Lawrence. During the following centuries, the hunting of the beaver spread to the Great Lakes, the Northwest Territories, and finally in the nineteenth century to the Pacific coast.

However, in the fur trade, European traders and native peoples were still getting mutual benefit. Because of the great distances and lack of knowledge of local conditions, Europeans were not able to organize the fur trade alone. They had to trust the skills and knowledge of native people, who knew the best hunting areas, animals, and traditional hunting techniques. A good example is the technology of transport, which in Europe was based on wheeled vehicles which were completely useless in forests. Instead traditional transport technology, like canoes, snowshoes, skis, and sledges, was vital in roadless wildernesses.

Whaling

The environmental history of whaling in the northern seas is a typical example of the extractive use of natural resources. From the middle of the nineteenth century onwards, steam-powered whaling boats penetrated further northwards into the waters of the Arctic archipelago. Markets for whale products remained good until the beginning of the twentieth century, when fossil oil and plastic began to substitute for materials made from whale oil and bones. Flexible whale bone was used as a material for umbrellas, hats, brushes, and corsets (Zaslow 1971, 249).

One whale species after another was killed until the whole industry became unprofitable. Extractive whaling expanded as a vicious circle from the Atlantic to the Arctic Ocean, causing whale populations to decline. The whale population in the fjords of Spitzbergen (founded in 1596) collapsed after only 30 years of intensive whaling (Bernes 1993, 15). At the end of the seventeenth century the eastern Atlantic and Greenland Sea became unprofitable for whaling (Rea 1976, 35; Hertz and Kapel 1986, 146).

From the middle of the nineteenth century, when steam-powered ships were introduced, whaling became much more intensive, and it spread even into arctic waters. In the Bering Sea, Ohota Sea, and Beaufort Sea, whole populations of bowhead whales collapsed within 60 years. The whale populations of Baffin Bay and Hudson Bay suffered from excessive hunting by the early twentieth century (Zaslow 1971, 257–58; Bockstocke 1984).

In the Canadian arctic archipelago whaling indirectly put a strain on other local animal populations. The whalers began to winter in the North to avoid expensive costs of travel from Europe. The result was that the wintering places developed into good marketplaces for fish and meat. During the winter months the crew of only one ship consumed about 150 caribou and also other game like musk oxen and walruses. Around wintering ships the populations of these animals began to decline. However, the most scarce resource was driftwood, used as fuel in these treeless areas (Zaslow 1971, 258). Extractive whaling broke the economic backbone of the Thule culture (1300–1750), rapidly changing the way of life and culture of the Inuits. They had to abandon their traditional permanent houses and villages and return to a nomadic way of life (Taylor 1971, 160–69).

In sum, the early phase of the developing European world-economy in the circumpolar zone, with its base in extractive use of animal resources, was not sustainable. The main threats to the northern native peoples were epidemics and alcohol closely associated with the extractive economy. Already the shift to European food and clothing was enough to break native resistance to disease (Tanner 1944, 458).

For example, measles, a relatively harmless children's disease in Europe, killed the whole of the Inuit population living on the Mackenzie delta at the beginning of the twentieth century. In 1910, there were only 120 people left from the original 2000 Inuits (Hargrave 1971, 194; Crowe 1974, 104–10). The area would have remained uninhabited in the absence of new Inuit migration from Alaska. In the same period, the epidemics spread by whalers killed half the Inuits living in Alaska and the whole population of Aleuts. The big villages of the Tlingit and Haida Indians suffered badly from smallpox epidemics. Only the Athabascans living in the interior were saved from this kind of catastrophe.

INDUSTRIALIZATION OF THE CIRCUMPOLAR NORTH

Beginnings of the Industrialization

Toward the end of the nineteenth century, the early commercial economy of the North was diversified by forestry, mining, and pulp and paper industries, initiating the new industrial era of northern development. Hydroelectric power and better traffic and transport communications were associated with more intensive economic development.

At first, the new industrial activities remained as enclaves—separate sectors—outside the traditional economy. However, it is analytically useful to distinguish industrialization based on "hard" and "soft" enclaves. Hard enclaves became typical in Canada and Russia, where extractive resource use began penetrating relatively untouched wildernesses by using modern technologies and demanding large-scale investments. In Scandinavian countries one can talk of industrial development based on "soft" enclaves. There the relatively stable agricultural population yielded a local labor reserve for the new industries and also some ready infrastructure, lowering quite remarkably the threshold for industrial development.

The sawmill industry opened up the southern parts of the circumpolar zone to the extractive timber trade with large-scale ecological impact. In Canada, the original British demand for timber was too heavily concentrated on the stocks of Scots pine. The logs were skimmed off the forests like cream from milk without paying any attention to the regrowth of the pinewoods. Already by the end of the 1890s, the sawmill industry was in trouble due to unsustainable logging. However, this crisis was attenuated by the advent of new groundwood and pulp mills dependent on fir forests. This new technology provided the basis for the large-scale penetration of the forest industry into the northern forests (Rea 1976, 40–41).

Initially, mining in the circumpolar zone was restricted to the most accessible regions. Because of the high risk and transportation costs involved, it was profitable to mine only the most valuable metals. The Klondike gold rush is a very good example of this. It began in the 1870s and was reaching its climax in 1896 when the richest gold deposit in Bonanza creek was found. In Canada, the major ore deposits in Sudbury (1993) and Cobolt (1903) were found when the construction of the transcontinental railway to the Pacific coast began.

For example, Sudbury, with the largest nickel deposit in the world, developed into one of the leading international mining centers. Mining, of course, had its adverse environmental effects. The forests around the mines were cut down and habitats were endangered by sulfur and arsenic emissions (Zaslow 1988, 158).

Hydroelectric power became the new energy source for industrial activities in the North. Traditionally, developers saw water power as a pure energy, a kind of white coal, with no or very small negative effects on the environment. At first, even conservationists had no criticism of such projects, because information about the ecological impacts of hydroelectric plants and man-made lakes was seriously lacking. The North was seen as an empty wilderness with plenty of room for nature and native groups. Besides, it was perhaps difficult to oppose this kind of impressive national effort, backed up by the leading politicians and businessmen (Massa 1985, 474).

The conservationists and native groups woke up only in the 1970s and began to campaign against the ecological and social costs of power plants and man-made lakes (Gill and Cooke 1974). Methods of social and environmental assessment of large technological projects only came into general use in the United States at the beginning of the 1970s and in Europe some years later. In 1971, the James Bay Development Project in northern Quebec was one of the first hydroelectric schemes where ecological and social costs had to be taken into account due to the increasing criticism from conservationists and Cree Indians (Massa 1985, 474–75).

Already in the 1960s, there was much discussion in the West on the possible ecological costs of the mega-projects the Soviet Union was planning. Among the most impressive were the plan to divert the rivers of Siberia which empty into the Arctic Ocean to the south and, it is hard to believe, a still more fanciful plan to warm up the arctic climate by damming up the Bering Strait (Cattle 1985, 485–96). It seems possible that these plans were only put forward for propaganda purposes to build the image of the Soviet Union as an almighty superpower and a dynamic industrial nation. The extractive mega-projects also exemplify the ruthless cultural interpretation of nature in the Soviet Union, the full consequences of which only came to light with the collapse of the Soviet empire (Susiluoto 1991, 328–29, 1995; Yanitsky 1993, 9–15).

World War II as a Turning Point

Northern Finland. World War II represents a turning point in the environmental history of the circumpolar zone as it rapidly raised the value of the zone's resources. In northern Finland, the so-called Continuation War (1941–1945) against the Soviet Union resulted in a new extractive industrialization program. With the loss of Karelia, Finland was deprived of its economically advanced core and about 10 percent of its productive capacity. The forest industry, the cornerstone of Finland's development, was the worst affected. One tenth of the country's forest resources, a quarter of the capacity of the sawmill industry, as well as many pulp mills, paper, plywood, and furniture factories were surrendered to the Soviet Union. The harbor of Vyborg-Uuras, once the biggest port in Fennoscandia for the overseas trade in processed wood products, was also lost.

After the war, Finland began to look north to compensate for the loss of Karelia and by the early 1950s Northern Finland was established as a resource periphery. In 1951 the report of the Committee for Industrialization was published, and a year later Urho Kekkonen, then prime minister, wrote an influential book entitled *Onko maallamme malttia vaurastua?* ("Can Our Country Wait to Prosper?") in which he suggested that the state should shoulder the responsibility for industrialization and resource development in Finnish Lapland, even if this were to entail a decline in the standard of living in other parts of Finland.

Kekkonen was a backwoods Finn, and as a chairman of the Agrarian Party (*Maalaisliitto*) he was interested in raising living standards in the countryside. Kekkonen's ideas were in sympathy with those of the Committee for Industrialization (1951) which had drawn up an industrial strategy for Northern Finland suggesting that the natural resources of Lapland should be utilized to accelerate industrialization (Komiteamietint 1951).

After the war, the value of the resources of the eastern and northern peripheries of Finland, which had been relatively insignificant in the 1920s and 1930s, began to grow rapidly. The area for economically profitable forestry expanded to the North and East. The first plans were published concerning the possible use of forest resource potentials of the river system emptying into the Arctic Ocean.

Hydroelectric power development assumed greater urgency with the onset of the serious energy crisis during and especially after the war, when fossil fuel imports ceased almost completely. The pulp and paper companies formed consortia for the construction of dams on the Oulu and Kemi Rivers where two-thirds of Finland's relatively small reserves of hydropower were situated.

Even traditional colonization activities were taken into consideration, because after the war the food supply was uncertain. During the war, politicians had promised land to servicemen. After the war Finland had to build almost 100,000 new farms for Karelian evacuees, war invalids, widows, and ex-servicemen. Some of the farms were established in the cold and remote areas of the East and the North.

The ecological costs of the new extractive strategy for industrializing northern Finland have been considerable. From a central European perspective, Finland probably still appears as a country of untouched forests. However, strictly speaking, virgin forests are very difficult to find any longer. The expansion of forestry has left marks particularly in northern and eastern Finland in large areas of seedlings, the draining and ploughing of swamps and forest lands, and in a growing network of roads.

Already in the 1960s, conflicts over land use between forestry, reindeer management, and tourism were becoming intense. The problem was exacerbated by large clearcutting operations carried out especially in the forests owned by the Department of Forestry, flouting sustainable yield forestry principles. The subsequent decrease in forest resources seriously undermined the expansion potential of the forest industry in the North.

The development of the Oulu and Kemi River hydropower schemes after World War II involved a serious reduction of the fish resources used by both the "salmon peasants" (*lohitalonpojat*) and the Lapps living on the upper reaches of the Kemi River. The first dam on the Kemi River, Isohaara, was built at the river mouth, putting an end to the salmon fishing which for centuries had marked the economic and cultural life of Lapland. Salmon and other anadromous fish species (whitefish, trout, lamprey) that used to spawn in the huge river disappeared all at once. Both rivers are now used nearly exclusively to provide energy for the country at large and the outside agencies supporting and controlling this development (Massa 1985).

As for agricultural colonization, the high climatic hazard factor in the East and North was almost completely ignored, and large numbers of the new so-called "cold farms" proved economically unviable. Consequently, settlers became increasingly dependent on wage work in forestry and civil engineering projects, especially as state support for the new farms turned out to be insufficient and as corn cultivation failed in the new fields.

The extractive industrialization program of the North and East of Finland led to a kind of ecological colonialism with little regard for the traditional industries or the conservation of resources.

The Soviet Union. In the Soviet Union, the importance of the natural resources in northern Russia and Siberia began to grow during Stalin's industrialization campaign in the 1930s. The labor force for the industrial activities in the North was either taken by force (prisoners, dissidents, religious fundamentalists) or by paying more or giving other benefits. During 1926–1956, the population of Siberia grew four- to fivefold (Eidlitz 1979; Armstrong, Rogers, and Rowley 1978, 45–53).

World War II and the subsequent Cold War were major incentives for resource mobilization in the Soviet North. For the Communist Party, the most important objective was to become an economic and military superpower in the struggle with the United States for world supremacy. Inside this highly centralized system, little thought was given to the increasing costs and risks of production. Since the collapse of the Soviet empire, this highly extractive system has been characterized as an "ecological crisis" (Susiluoto 1991, 326–27; Lemechev 1991; Peterson 1993).

Canada. During and after the war, the economic center of gravity in Canada also shifted to the North and almost the whole of northern Canada, from Labrador to Alaska, came under the influence of the various war projects (Zaslow 1988, 209). Several arctic airfields, new telecommunications networks, and the Alaskan highway were built especially for transporting casualties and military technology. The United States and Canada together built a network of weather and radar stations along the northern frontiers of Canada. Even technological development during the war—freighters, air photography, and nuclear technology—paved the way for resource exploration ever further north (Armstrong, Rogers, and Rowley 1978, 87–88; Rea 1976, 55).

As the world's leading economic superpower, the United States' postwar growth rapidly increased its consumption of raw materials, further exposing the great shortcomings of its domestic raw-material supplies. In these circumstances, the position of Canada as a raw material producer for its southern neighbor strengthened notably after the war. The demand for resources grew also in Europe, where postwar reconstruction needed plenty of raw materials.

The guidelines for the new situation were set out in the so-called Paley report, Resources for Freedom (1952), financed by the U.S. government. In the Paley report, 22 "key materials" and their potential supply areas were identified. Canada was seen as the most secure source for 13 strategically important raw materials (nickel, copper, lead, zinc, asbestos, iron ore, sulfur, titanium, cobalt, oil, natural gas, paper, and aluminum). The result was that Canada was integrated more and more within the sphere of influence of the United States as a kind of "specialised resource producer" (Hayter 1985, 439; Wallace 1985, 480).

However, U. S. "resource capitalism" (as the critics called it) proceeded in harmony with the aims of the government of Canada, which was trying to attract new investments to the country. A good example is Prime Minister John Diefenbaker's program, Roads to Resources (1950–1970), aimed at opening up the northern parts of the provinces (the so-called Middle North) for development. The most important instrument of the program was infrastructure: new railways and roads (Zaslow 1988, 253). Later, conservationists and native people began to question the ruthless extraction of resources.

There is only fragmentary information about the ecological costs of the intensified extractive resource strategy in Canada after the war. However, the traditional cut-and-run forestry (or forest mining) with its adverse environmental effects has continued in Canada in the absence of any strong political pressure to shift to more sustainable forestry (Swift 1983; Marchak 1983; MacKay 1985).

Overall, the effect of World War II was a relative shift in the center of gravity of the world-economy toward the North. The growing significance of the northern zone as a resource periphery of the world-economy is a very important change. Southern parts of the northern zone were technologically and culturally opened up for extractive industrialization. The ecological costs of development were almost totally ignored during the decades of post-war reconstruction and intensive economic growth.

GLOBAL RISK ECONOMY IN THE CIRCUMPOLAR NORTH

Ilmari Hustich, a Finnish geographer specializing in northern areas, wrote in 1973 the following ecologically enlightened vision for the northern zone:

> I believe, however, that in the near future the Middle North to a larger extent will be needed and used for normal and particularly more careful colonization. It seems probable, however fanciful it might sound today, that the area under cultivation will again soon increase and that a new expansion of the inhabited area in the North will take place. For instance, the new farmland abandonment policy of Sweden and Finland is likely to look rather silly in the near future. There will also be an increased necessity to grow vegetables in regions where pure water and cleaner air will guarantee a better product. . . . In the developed countries of the northern hemisphere people usually, mentally speaking, look southwards only. And in this direction there are certainly reasons to be pessimistic; in the southern latitudes the problems of mankind are accumulating. I have tried to show that a slightly, but just slightly more optimistic view can be found if we look to the North, to the still more or less empty, partly unexplored and as yet not so destroyed Arctic, Subarctic and Middle North regions.

Hustich saw very early the new role of the circumpolar zone in a risk civilization threatened by more and more serious environmental problems. However, we can see more clearly that this vision was based on a myth of the virgin North lying beyond the evil world. First, social problems, especially unemployment, have continuously become more acute and the agricultural policy of the European Union is now seriously aggravating the situation. If nothing is done, at least the northern and eastern peripheries of Finland will be deserted by the end of the twentieth century. In Finland, the so-called green wave has not turned out to be very strong despite some "eco-experiments," perhaps because the structure of Finnish communities is already relatively dispersed.

Secondly, a growing consciousness is developing now that global environmental problems are putting the possibilities of finding effective shelters against the evil world in a new light. Generally speaking, the new kind of "geographical insecurity" that is slowly emerging will create a new kind of global consciousness hopefully.

In what follows, I will discuss only four important major trends—oil and gas development, chemical pollution, forest cuttings, and nuclear risks—connected with the development of the global risk economy in the circumpolar zone. All of these megatrends have already had very profound cultural and ecological impacts in the global North. However, in assessing the importance of these trends, one must remember that even today the northern zone is one of the cleanest regions in the world.

Oil and Gas

In global environmental history there are two very important megatrends. The first is the shift to farming and animal husbandry, which are not very important in the circumpolar north excluding northern Fennoscandia. The second is the shift from the solar age to the fossil age, which is still going on. Over the past 150 years, humans have become the first and only species to have transcended the system of renewable solar energy by exploiting nonrenewable fossil fuels (coal, oil, natural gas, etc.) stored in the earth's crust over millions of years. This shift from "the solar age to the fossil age" was a necessary prerequisite for industrialization, economic growth, and the welfare state (Daly 1983, 342). It is also a direct cause of many recently recognized global forms of environmental damage, such as acid rain and the greenhouse effect (Sieferle 1990, 13).

Seen in historical perspective, the shift to a new fossil-fuel-based energy system is a very new phenomenon. For example, in Germany, even in 1939, there were still three million horses in use. In Finland, where logs were traditionally transported from forests to floatways by horse power, the number of horses reached its maximum as late as the end of the 1940s.

In recent decades, the mobilization of oil and gas has overshadowed almost all other resource projects in the north. In 1959, the second largest gas field found in Slochtern (Netherlands) was a very important starting point for oil and gas development in the North Sea and even in other northern seas. Gas was found in a rock layer that extends from the coast of the Netherlands toward the continental shelves of England and Norway. Another very important factor working behind the expansion of oil and gas development in the northern seas was the offshore technology developed in the 1960s in Lake Maracaibo and the Gulf of Mexico.

However, it was the discovery of the Ekofisk oil and gas field in 1969 that really started up the fossil fuel boom in the North Sea. Exploration was intensified and several other fields were found along the continental shelves of Norway and Great Britain. Both countries developed rapidly into very important oil producers. In 1994, for example, Norway got 30 percent of its import revenues from North Sea oil.

In Alaska in the 1950s and 1960s big oil deposits were found on the Kenai Peninsula and Cook Bay. However, a very important turning point came in 1968 when the Atlantic Richfield Company (ARCO) and Humble Oil and Refining Company (now Exxon) made the biggest single oil discovery in northern America on Prudhoe Bay. In 1977, oil began to flow to world markets through the 1300-kilometer-long pipeline from Prudhoe Bay to Valdez. At the beginning of the 1990s Alaska already produced 25 percent of the oil in the United States and 10 percent of its total consumption (Jumppanen 1990, 74).

This discovery launched a kind of oil rush in Alaska and even in other parts of the circumpolar zone. At the beginning of the 1970s exploratory drillings were begun in the Beaufort Sea and off the coast of Newfoundland in Canada, with the result that many promising discoveries were made in the Mackenzie

Delta, Svedrup Basin, and the continental shelf of Newfoundland–Labrador (Hibernia oil field). Disputes between the federal government and the province on the rights of possession and, after 1986, the lower price of oil have postponed production, however (Armstrong, Rogers, and Rowley 1978, 93–99; Staveley 1982, 244–46; Jumppanen 1990, 74).

In the Soviet Union, the first oil rush began in 1953 when one of the biggest oil deposits in the world was found by chance in the West-Siberian Lowland between the Ob and Jenisei Rivers. Production began there in 1963–1964. In the 1970s and 1980s, huge gas fields—Urengoy, Jamburg, Bovanenko, etc.—were discovered in Western Siberia and the Jamali Peninsula. It is estimated that a third of the known oil reserves in the world are in these fields (Jumppanen 1990, 73). In this region, there are already two cities with about 100,000 inhabitants. Oil and gas production is also going on in the Petsora River delta and at some other smaller deposits (Armstrong, Rogers, and Rowley 1978, 31–36; Goldman 1980, 34–35).

Despite serious difficulties in production, Russia is still the world's leading oil and gas producer, and northwestern Siberia produces 78 percent of the oil and 84 percent of the gas in Russia. Native groups living in this area have begun to organize themselves to defend their rights against the pressures of the oil and gas industry (Chance and Andreeva 1995). It is highly probable that the oil and gas industries will expand in the near future to the continental shelves of the Arctic Ocean, where exploratory drilling is already going on (Clarke 1991, 121). A huge gas deposit (Shtochkmanovskoy) with up to 4000 billion cubic meters of gas found in 1988 in the eastern part of the Barents Sea is one of the most promising. The Scandinavian countries and Russia formed The Barents Sea Commission in order to organize cooperation for exploiting this huge deposit.

Oil and gas development has already caused significant local environmental damage. One of the biggest disasters was in 1991, when 42 million liters from the oil tanker Exxon Valdez polluted the whole of the Prince William Sound and the coastlines of Alaska Bay as far as the Aleutian Islands. In the delta of the Ob River in Siberia, a leak from a cracked pipeline caused serious environmental damage. In the fall of 1994 the same kind of accident happened in the Komi region—an oil leak of 70,000–80,000 tons, which seriously polluted local ecosystems. There is also information about the very careless transportation of oil on the Arctic Ocean that is littering and polluting the whole of the Siberian coastline (Vitebsky 1990; Pentikäinen 1990; Rogingo 1992, 146).

During the past 20 years oil and gas have developed as dominant industries in the northern zone, which has about 15 percent of the world's known reserves of oil and almost 30 percent of world gas reserves.

Today, Russia is still self-sufficient in oil production, whereas the United States already imports 40 percent of its consumed oil. In any case, both countries in the near future will have to fall back on their arctic reserves if they do not want to become dependent on the outside, in this case mainly the reserves of the Middle East.

Oil and gas exploration will probably expand during the next decade to the continental shelves and coasts of seas in the North. It is possible that gas production, for ecological reasons, will take precedence over oil-based development. It is estimated that gas reserves on the Jamali Peninsula, the Barents Sea, the north slope of Alaska, and the Mackenzie delta will be in production already in the year 2000 (Jumppanen 1990, 71–77). The collapse of the Soviet Union has removed the high ideological barrier to arctic economic cooperation. Already Siberian hydrocarbon potential is opening up to western companies. It is likely that the serious economic crisis in Russia will speed up the use of hydrocarbons in the North.

Chemical Pollution

The so-called arctic haze is one example of the increasing chemical pollution of the atmosphere reaching as far as the northern zone. The first references to arctic haze are from the 1950s, when it was considered a normal phenomenon in the north. In the 1970s, more accurate research revealed that the arctic haze was in fact acid gas consisting of several sulfates and aerosols. It is most common from December to April when southern air currents bring pollution from the industrial centers. Arctic haze is a typical example of global pollution, although some writers have suggested that the heavily polluting smelters and wood processing industries of the Russian North might be responsible for this kind of pollution. Thus far, there is no consensus about the possible environmental effects of arctic haze (Pryde 1991, 21; Russia, State of Environment 1994).

The risk of acidification is greatest in the boreal forests of Fennoscandia, northwestern Russia, and eastern Canada, where the moraine soils are vulnerable to increasing acidification and pollution from nearby industrial centers. The biggest local polluters are the industrial plants on the Kola Peninsula, which emit about 0.8 million tons of sulfur oxide annually, and in the region of Norilsk, with over 1.9 million tons of sulfur oxide emissions (Nenonen 1991). Rogingo (1992, 14) argues that 60 percent of the population living in Kola are suffering from health problems related to industrial pollution.

The "greenhouse effect" might have the most influence in the northern zone. The history of the ice age shows that even relatively small atmospheric changes cause dramatic changes in ice masses. These changes have also had significant cultural impacts, as the colonization history of Nordic countries is clearly revealing. However, climatic models cannot predict accurately what will happen in the North, although most climatologists believe that increasing carbon dioxide discharges will raise the temperature in the North as well. The result could be that ice masses in the North will begin to melt, raising sea levels and causing serious problems all over the globe (Pryde 1991, 30).

The possible thinning of the ozone layer over the North Pole is one of the most recent global environmental threats. In Canada, and even in Germany, the public has been advised to avoid unprotected exposure to sunlight.

Forest Cutting

The northern boreal forest, which contains one-third of the world's timber, is the most important raw-material source for the international pulp and paper industry.

Most of this trade links the countries of the northern boreal zone to consumers in Western Europe, the United States, and Japan. Within the northern boreal zone, the center of gravity of forest use is gradually shifting from Scandinavia and Eastern Canada to Russia and Western Canada. For example, in Canada two big pulp and paper plants have quite recently been built in northern Alberta, and in British Columbia forest companies have been permitted to cut almost all of the old growth forests.

International forest companies have turned their attention especially to Siberian forests, after partly relinquishing their former positions in the tropical forests, due to the declining supply of tropical timber and the sharp criticism of ecological conservationists. At the same time, Russia has opened up Siberian forests to international bidding by forest companies because it desperately needs new investments in order to overcome its deep economic difficulties. The companies have been encouraged to establish joint ventures with Russian partners in return for generous timber concessions of hundreds of hectares of forest land. In the Russian Far East alone, 66 joint ventures have recently been established (Petrof 1992; Russia, State of Environment 1994, 143–45).

The conservationists are afraid that Siberia is going the same way as the Amazon as far as extractive forest use is concerned. The assessment of this criticism is difficult because there is no consensus on the definition of the concept "old forest." Forest companies, of course, support the economic definition of an old forest as forest that consists of dead and decaying trees that have to be cut down periodically if we want to follow the directives of sustainable forestry. The definition of the environmental organizations emphasizes the biological and ecological value of the old forests. The forests have to have a long history of natural succession to be called old, and dead and decaying trees are indicators of high biodiversity and good quality (Wahlström, Hallanaro, and Manninen 1996, 79–80).

In any case, extractive forestry is ready to take over even the northern zone of Russia. For example, Prime Minister Viktor Chernomyrdin decided to start, in May 1995, annual cutting of a million cubic meters in the 900 kilometer forest zone between Finland and Russia. Many Finnish and Swedish forest companies were interested in buying timber from this area. However, the Russian Academy of Sciences began to criticize this plan because intensive cutting would destroy the relatively untouched forests in the frontier zone in five years. However, Mikhail Feshenko, Karelian Minister of the Environment, decided to restrict the cutting to 200,000 cubic meters per year, when the Russian Greenpeace and the Socioecological Union also launched a campaign against the cutting (*Taiga News* 1995).

South Korea, having lost its own forests during the Korean War, has been especially interested in utilizing timber from the forests of Siberia and the Russian Far East, but many American and Japanese companies also are involved. Finnish and Swedish firms have not been very active in these parts of the world, but, of course, Nordic forest technology is widely used in these projects (Rosencrantz and Scott 1992).

The history of forest use in the circumpolar zone is mostly extractive. However, there are still large so-called primeval forests in the North, which have remained outside the inhabited area and the world economy. From the point of view of the global risk civilization, it is easy to find strong arguments for special conservation measures in these forests. The forests of the northern zone are producing an important part of the oxygen for the world, and they also absorb carbon dioxide from the atmosphere. At the same time, however, the importance of the northern forests to the world economy is growing almost as rapidly. In the near future, the possibilities of deeper conflicts between producers and conservationists will certainly increase in the North.

Radioactivity Risks

Radioactivity risks in the North are perhaps the clearest example of the global risk economy. These risks are heavily concentrated in Russia, which has to shoulder the legacy of the careless handling of nuclear waste in the former Soviet Union. In the Soviet Union, the military status of a nuclear superpower was won at the cost of creeping nuclear catastrophe. Nuclear testing, nuclear submarines, and the dumping of nuclear wastes into the seas are the most obvious examples of these radioactive risks.

The United States carried out some nuclear tests on the Alaskan Amchitka Island, the details or consequences of which are not at all known. The Soviet Union made its largest nuclear tests on the arctic archipelago of Novaya Zemlya. From 1955 to 1990, a total of 132 nuclear tests were carried out on Novaya Zemlya, 90 of which were above ground, before atmospheric testing was forbidden by international agreement in 1963. The nuclear pollution from the tests in Novaya Zemlya has spread all over the northern zone. For example, the concentration of cesium-137 in milk produced in Finnish Lapland was 20 times higher than in southern Finland (Jokelainen 1965). However, subterranean tests have continued on Novaya Zemlya with the result, according to some Norwegian observers, that they have torn "craters" in the bedrock and permafrost of the island. Through the craters radioactive substances could leak into groundwater and the sea (Gizewski 1995).

Russian nuclear submarines are heavily concentrated in Murmansk on the coast of the Arctic Ocean and in Severodvinsk on the coast of the White Sea. According to Bellona's report, the density of nuclear reactors is the largest in the world on the Kola Peninsula: 240 nuclear reactors (Nilsen, Kudrik, and Nikitin 1996). Most of them are installed in nuclear submarines or cruisers of the Russian northern fleet, which has as many as 18 percent of all the reactors in the world. In addition, a shipping company in Murmansk owns eight nuclear icebreakers and one nuclear tanker. In a very risky nuclear power plant in Polarnyi Zorn there are still four reactors. Between 1945 and 1988 American or Russian nuclear submarines or cruisers suffered 20 serious nuclear accidents. Even the reactors in sunken ships have to be under constant control because of the risk of leaking nuclear pollution (Gizewski 1995, 28).

The third problem is the dumping of nuclear waste into the Arctic Ocean. During the past 30 years, the Russians have poured large quantities of radioactive waste into the Barents and Kara Seas. In 1993, President Boris Yeltsin appointed Andrei Jablokov's committee to collect information on the problem. According to the Committee, there are 11,000 barrels of solid or liquid radioactive materials, including 16 reactors from nuclear submarines, at the bottom of the sea off the eastern coast of Novaya Zemlya. In seven of them, even the fuel stacks have not been removed.

However, it is impossible to assess accurately the risk of the nuclear waste in the Russian North. Because neutral research is lacking, various rumors about cancers are widely circulating, some so horrendous that researchers suspect that Moscow is spreading them purposely in order to get more western money (Gizewski 1995, 32–33).

There are some signs that a more active environmental policy is slowly developing even in the circumpolar zone. In 1991, the Finnish government made a suggestion for an environmental conservation strategy that has created an important discussion forum for the respective northern countries. In 1996, it was decided that a treatment plant for nuclear wastes will be built in Murmansk with Russian, American, and Norwegian money (Mustonen 1996). Unfortunately, many of the environmental risks are still developing beyond the reach of the control mechanisms established by the nations involved and international organizations.

NATIVE LAND CLAIMS

In the 1960s, governments could still give large natural resource concessions in the North to multinational companies without asking permission from the local peoples. Now the situation is more complex. The native people have started to organize themselves and to oppose the projects where their rights to land and culture have been set aside. This development started in Alaska after 1959 when Alaska gained statehood. The state government got the right to separate from federal lands the economically most important areas for itself. To the civil servants' surprise, this development faced resistance from the Inuits, Indians, and Aleuts living outside the cities.

The native people of Alaska were organized in the mid 1960s as a federation (The Alaska Federation of Natives), which started to fight for the rights of aboriginal people. Native land claims began to pile up on the desks of government officials and eventually covered more than the whole area of Alaska as some of them spilled over into Canada.

The discovery of the Prudhoe Bay oil deposit in 1968 suddenly changed the whole situation. The oil was found on state land but the Federation of Natives had not given up their claims to the same land. This process and, in fact, the whole question of land claims in Alaska was settled in 1971 by the Alaska Native Settlement Act under which native people gave up their rights to their ancient lands.

According to the act, 12 culturally homogeneous native regions were established, each run by a regional corporation controlling the division of lands and the use of money. In addition, new village corporations were founded. Every Alaskan native had to prove that s/he was at least one-quarter Inuit, Indian, or Aleutian. After that s/he had to join a certain regional or village corporation to get compensation for his/her native rights (Cooley 1984, 20).

Some of the new regional or village corporations and their managers have become quite wealthy from the monetary compensation for the use of their lands. But the interests of the traditional hunters and the new corporations can conflict. Hunters are interested in preserving the environment for hunting and fishing. The managers are, of course, mainly interested in reaping as much benefit as possible from the monetary compensation, and often this means that they want to invest in multinational companies outside Alaska (Cooley 1984, 22).

The native peoples of Alaska have sold their birthright to their land for a relatively high price. However, this has mostly happened at the cost of their traditional way of life. The Alaskan case was very important even for native peoples living in other parts of the circumpolar zone. In Fennoscandia, the situation of the Sami (Lapps) is quite complex because the colonization based on farming and animal husbandry, which began in the fifteenth century, has reached the coast of the Arctic Ocean, so that the relations of colonists and native peoples are profoundly intertwined. In Siberia, the Yakuts especially have highlighted their independence, even establishing an autonomous republic called Saha. In Canada, the land claims of the native peoples cover almost all of the Canadian North, including the northern parts of the provinces and both territories. In 1973 the first lawsuit was filed by the Nishga Indians in British Columbia. The Indians lost the case, but as a consequence Canada has had to change its policy toward the northern native peoples (Henriksson 1989). In 1998 the Nishga achieved a lucrative settlement.

The James Bay and Northern Quebec Agreement of 1976 is a good example of the new policy. Under this agreement, the Cree Indians and Inuits gave up most of their traditional lands for monetary compensation and relatively modest land areas (Quebec, The James Bay and Northern Quebec Agreement 1976; Richardson 1972). Similar agreements have been reached in other parts of Canada as well. For example, in 1993, the Nunavut Land Claim Agreement was signed with the Inuits.

Related environmental conflicts are now occurring in Siberia and northern Russia. For example, a Russian–Korean joint venture, Svetlaya, backed by the Korean multinational Hyundai, lost a lawsuit in the Russian High Court against an environmental organization (the Socioecological Association) and two native peoples (the Udegaits and the Nanaits). It had to give up a concession of 300,000 hectares of forest land in the valley of the Bikini River, which the regional government in Primorsky had granted it. This decision was based on President Boris Yeltsin's statute in 1992 according to which native people have to give permission for projects that clearly threaten their traditional rights. There are also other land claim issues in Russia based on this statute (Winestock 1992).

These experiences, especially those in Alaska and Canada, show that native organizations are mobilizing all means to get benefits from the states and firms who are keen on developing their traditional lands. Often the monetary compensation is invested in modern technologies (e.g., in snowmobiles, motorboats, and luxuries), not in "green" or alternative investments as some conservationists had hoped. Ecological arguments are used as a part of native ideology, but the "ecological" way of life lives on only in cultural traditions which are rapidly being marginalized with the decline of self-sufficient economies.

CONCLUSION

Before Europeanization, the northern circumpolar zone was inhabited by very small groups of arctic and subarctic peoples. At first they had only casual contacts, but from the Middle Ages onward more permanent trade contacts were established with the developed cultures in the South. Their way of life was still dominated by the local natural environment and their own cultural traditions.

From the sixteenth century on, European culture spread from the southern centers to the northern zone until the whole area was under its influence. Now it is no longer possible to talk about Europeanization but only about the penetration of multinational industrial cultures into the north. Investment is not coming only from European or American companies but also from Japan, Korea, and other financial centers of the Far East.

The Europeanization of the northern zone is very easy to defend by referring, for example, to economic growth and the higher material welfare of the native peoples living there today. However, there are two basic omissions in this view, the significance of which will be seen only in the future. First, the penetration of European culture among the native peoples has either totally destroyed their cultures or has caused very difficult social problems like alcoholism and unemployment. These traditional northern cultures can still be seen in remote regions of Russia, but here too they are being opened for more intensive economic exploitation.

Second, the industrial extractive economy seems to have relatively sharp teeth in the northern zone because resistance against the developers has traditionally been quite low. Right now the northern zone is on the brink of new economic and ecological change. The zone is now part of the global risk economy, the development of which seems to escape national or international control attempts. The insecure material and energy situation in the global risk economy is manifest in the northern zone in the form of more and more resource extraction projects.

The value of the northern zone is rising again, not as an ecological hiding place or reserve area as was dreamt of a few decades ago, but as a raw material source of timber, ores, and especially fossil fuels. Multinational companies and "circumpolar" governments are waiting impatiently to tap the huge natural resource potentials of the circumpolar north.

International environmental policy is facing big challenges in the new resource situation in the North. The conflicts between developers, conservationists, and native peoples will increase in the future.

The northern resource projects should no longer be based on the "progressive" thinking originating in previous centuries but on the holistic view in which not only economic but also ecological, physical, social, and cultural aspects are considered. Far from simply being the last resource frontier for industrial countries, the northern zone is also home for many native peoples and, in addition, an increasingly important part of the biosphere. The future of the northern zone should thus not only be based on natural science, but on new social science as well.

NOTES

1. The references for this paper come from general geographical, historical, and anthropological research on the circumpolar north. Part of the chapter is drawn from a recent environmental history of Finnish Lapland, thus far published only in Finnish (Massa 1994). I have also used the Internet to bring information up to date; for example, the most recent reports of environmental organizations, like Taiga Rescue Network, Greenpeace, and Bellona, are available on the Internet.

2. Interestingly, there has been in Finland a somewhat older debate on "risk economy" that, however, was not based on Beck's ideas but instead originated from a much older German concept of *Raubwirtschaft*, as presented by Ernst Friedrich (1904) at the beginning of the twentieth century. This concept has been variously translated as the destructive, exploitative, or pillage economy (Raumolin 1984; Massa 1994; Massa 1995). The term Raubwirtschaft has been used to describe how the ecological adaptation of industrialized societies is based on overexploitation of nature and its resources, disregarding the interests of future life and future needs. The concept has been associated only with industrial societies, emphasizing the enormous weight of industrialization in the environmental history of the world.

REFERENCES

Armstrong, Terence. 1958. *The Russians in the Arctic.* London: Methuen.

Armstrong, Terence. 1965. *Russian Settlement in the North.* Cambridge, UK: Cambridge University Press.

Armstrong, Terence, George Rogers, and Graham Rowley. 1978. *The Circumpolar North: A Political and Economic Geography of the Arctic and Sub-Arctic.* London: Methuen.

Attman, Arthur. 1973. *The Russian and Polish Markets in International Trade 1500–1600.* Göteborg: Publications of the Institute of Economic History of Gothenburg University.

Beck, Ulrich. 1986. *Risikogesellschaft: auf dem Weg in eine andere Moderne.* Frankfurt am Main: Suhrkamp.

Bernes, Claes. 1993. *Pohjoismaiden ympäristö—tila, kehitys ja uhat.* Nord 13.

Bockstocke, John. 1984. From Davis Strait to Bering Strait: The Arrival of the Commercial Whaling Fleet in Northern America's Western Arctic. *Arctic* 37: 4, 528–32.

Braudel, Fernand. 1984. *The Perspective of the World.* Vol. 3 of *Civilization and Capitalism: 15th–18th Century.* New York: Harper & Row.

Bunker, Stephen G. [1985] 1988. *Underdeveloping the Amazon: Extraction, Unequal Exchange and the Failure of the Modern State.* Chicago and London: University of Chicago Press.

Cattle, Howard. 1985. Diverting Soviet Rivers: Some Possible Repercussions for the Arctic Ocean. *Polar Record* 22: 485–98.

Chance, Norman A., and Elena N. Andreeva. 1995. Sustainability, Equity, and Natural Resource Development in Northwest Siberia and Arctic Alaska. *Human Ecology* 23: 217–40.

Clarke, James W. 1991. Oil and Gas Resources in the Offshore Soviet Arctic. In *The Soviet Maritime Arctic,* ed. Lawson W. Brigham, 108–22. London: Belhaven Press.

Cooley, Richard A. 1984. Evolution of Alaska Land Policy. In *Alaskan Resources Development: Issues of 1980s,* ed. Thomas A. Morehouse, 13–49. Boulder, CO: Westview Press.

Crosby, Alfred. 1987. *Ecological Imperialism: The Biological Expansion of Europe, 900–1900.* Cambridge, UK: Cambridge University Press.

Crowe, Keith J. 1974. *A History of the Original Peoples of Northern Canada.* Montreal and London: McGill–Queen's University Press, Arctic Institute of North America.

Daly, Herman E. 1983. The Steady-State Economy. In *Annotated Reader in Environmental Planning and Management,* ed. Timothy O'Riordan and R. Kerry Turner, 237–53. Oxford, UK: Pergamon Press.

Eidlitz, Kerstin. 1979. Revolutionen i norr. Om sovjetetnografi och minoritetspolitik. Uppsala universitet, Kulturantropologiska institutionen. Uppsala.

Friedrich, Ernst. 1904. Wesen und geographische Verbreitung der Raubwirtschaft. *Petermanns Mitteilungen* 50: 68–79, 92–95.

Gill, D., and A. D. Cooke. 1974. Controversies over Hydroelectric Development in Sub-Arctic Canada. *Polar Record* 17: 109–27.

Gizewski, Peter. 1995. Military Activity and Environmental Security: The Case of Radioactivity in the Arctic. In *Environmental Security and Quality After Communism: Eastern Europe and the Soviet Successor States,* ed. Joan DeBardeleben and John Hannigan, 25–41. Boulder, CO: Westview Press.

Hamelin, Louis Edmund. 1978. *Canadian Nordicity*. Montreal: Harvest House.

Hargrave, M. R. 1971. Changing Settlement Patterns amongst the Mackenzie Eskimos of the Canadian Western Arctic. In *Canada's Changing North,* 187–98. Carleton Library No. 55. Toronto: McClelland & Stewart.

Hayter, Roger. 1985. The Evolution and Structure of the Canadian Forest Product Sector: An Assessment of the Role of Foreign Ownership and Control. In *Natural Resources and Problems of Staples-Based Industrialization in Finland and Canada,* ed. Jussi Raumolin. *Fennia* 163(2): 439–50.

Henriksson, Markku. 1989. Kruunun suojeluksessa–intiaaniasianhoitoa Kanadassa. *Historiallinen Aikakauskirja* (1): 36–47.

Hertz, O., and F. O. Kapel. 1986. Commercial and Subsistence Hunting of Marine Mammals. *Ambio* 15(3): 144–51.

Hustich, Ilmari. 1972. Den arktiska, subarktiska och boreala regionens folkmängd. *Terra* 84(3): 181–90.

Hustich, Ilmari. 1973. The Arctic and the Subarctic: Middle North-Regions and their Future. *International Council of Science Unions Bulletin* 5–19.

Innis, Harold. [1930] 1956. *The Fur Trade in Canada: An Introduction to Canadian Economic History.* Toronto: University of Toronto Press.

Innis, Harold. [1940] 1954. *The Cod Fisheries: The History of an International Economy.* Toronto: University of Toronto Press.

Jokelainen, Aili. 1965. Diet of Finnish Lapps and its Cesium-137 and Potassium Contents. *Acta Agralia Fennica* 103: 1–140.

Jumppanen, Pauli. 1990. Environmental Aspects of the Exploitation of Arctic Oil and Gas Reserves. In *Arctic Environmental Problems,* ed. Lassi Heininen, 70–84. Tampere Peace Research Institute Occasional Papers 41. Tampere.

Komiteamietint. 1951. Committee on Industrialization Annual Report 12.

Lemechev, Mikhail. 1991. *Désastre écoloqique en U.R.S.S.* Paris: Éditions Sang de la Terre.

MacKay, Donald. 1985. *Heritage Lost: The Crisis in Canada's Forests.* Toronto: Macmillan Canada.

Marchak, Patricia. 1983. *Green Gold: The Forest Industry in British Columbia.* Vancouver: University of British Columbia Press.

Massa, Ilmo. 1985. Hydroelectricity and Development in Northern Finland and Northern Quebec. *Fennia* 163(2): 465–77.

Massa, Ilmo. 1994. *Pohjoinen luonnonvalloitus: Suunnistus ympäristöhistoriaan Lapissa ja Suomessa.* Helsinki: Gaudeamus.

Massa, Ilmo. 1995. Historical Approach to Environmental Sociology. *Innovation in Social Science Research* 8(3): 261–74.

Morehouse, Thomas A. 1984. Introduction. In *Alaskan Resources Development: Issues of the 1980s,* ed. Thomas A. Morehouse, 1–12. Boulder, CO: Westview Press.

Mustonen, Sirpa. 1996. Arktisessa ympäristönsuojelussa edetään pienin askelin. *Ympäristö* 10: 2.

Nenonen, M. 1991. Report on Acidification in the Arctic Countries. Man-made Acidification in a World of Natural Extremes. In *The State of the Arctic Environment,* 7–81. Arktisen keskuksen julkaisuja 2. Rovaniemi.

Nilsen, Thomas, Igor Kudrik, and Aleksander Nikitin. 1996. The Russian Northern Fleet: Sources of Radioactive Contamination. Bellona Report 1.

Pentikäinen, Juha. 1990. Maakaasutuotanto tuhoaa Siperian tundran ja kulttuurit. *Yliopisto* 26: 8–9.

Peterson, D. J. 1993. *Troubled Lands: The Legacy of Soviet Environmental Destruction.* Cambridge, UK: Cambridge University Press.

Petrof, Divish. 1992. Siberian Forests under Threat. *The Ecologist* 22(6): 267–70.

Pryde, Philip R. 1991. *Environmental Management in the Soviet Union.* Cambridge, UK: Cambridge University Press.

Quebec, Canada. 1976. *The James Bay and Northern Québec Agreement.* Montreal: Editeur officiel du Quebec.

Raumolin, Jussi. 1984. L'homme et la déstruction des resources naturelles: la Raubwirtschaft au tournant du siècle. *Annales E.S.C.* 39: 788–819.

Rea, K. J. 1976. *The Political Economy of Northern Development.* Background Study 36. Ottawa: Science Council of Canada.

Richardson, Boyce. 1972. *James Bay: The Plot to Drown the North Woods.* Toronto: Sierra Club.

Rogingo, Alexei. 1992. Conflict between Environment and Development in the Soviet Arctic. In *Vulnerable Arctic,* ed. Jyrki Kakonen, 144–54. Research Reports 47. Tampere: Tampere Peace Research Institute.

Rosencrantz, Amin, and Anthony Scott. 1992. Siberia's Threatened Forests. *Nature* 355 (January 23): 293–94.

Russia. 1994. State of Environment of the Russian Federation 1993. National Report. Moscow: Ministry for Environment Protection and Natural Resources.

Sieferle, R. P. 1990. The Energy System: A Basic Concept of Environmental History. In *The Silent Countdown,* ed. P. Brimblecombe and C. Pfister. Berlin: Springer-Verlag.

Susiluoto, Ilmari. 1991. Yhteiskunnallinen ympäristökeskustelu Neuvostoliitossa. In *Ympäristökysymys: Ympäristöuhkien haaste yhteiskunnalle,* ed. Ilmo Massa and Rauno Sairinen, 326–48. Helsinki: Gaudeamus.

Swift, Jamie. 1983. *Cut and Run: The Assault on Canada's Forests.* Toronto: Between the Lines.

Taiga News. 1995.

Tanner, Väinö. 1944. Outlines of the Geography, Life and Customs of Newfoundland-Labrador. *Acta Geographica* 8(1).

Taylor, W. E., Jr. 1971. The Fragments of Eskimo Prehistory. In *Canada's Changing North,* 160–69. Carleton Library No. 55. Toronto: McClelland & Stewart.

U. S. Congress. 1952. House of Representatives. *Resources for Freedom (The Paley Report).* 82nd Congress, 2nd Session. Document 527. Washington, DC.

Vitebsky, Piers. 1990. Gas, Environmentalism and Native Anxieties in the Soviet Arctic: The Case of Yamal Peninsula. *Polar Record* 26: 156, 19–26.

Wahlström, Erik, Eeva-Liisa Hallanaro, and Sanni Manninen. 1996. *Suomen ympäristö tulevaisuus.* Helsinki: Edita and Suomen ympäristökeskus.

Wallace, Clement. 1985. Labour in Exposed Sectors: Canada's Resource Economy. In *Resources Exploitation and Problem of Staples-Based Industrialization in Finland and Canada,* ed. Jussi Raumolin. Fennia 162(2): 479–88.

Winestock, Geoff. 1992. Russian Supreme Court Rules Against Hyundai Logging Operation. *Moscow Times,* November 30.

Wolf, Eric. 1982. *Europe and the People Without History.* Berkeley: University of California Press.

Yanitsky, Oleg. 1993. *Russian Environmentalism: Leading Figures, Facts, Opinions.* Moscow: Mezhdunarodnyje Otnoshenija Publishing House.

Zaslow, Morris. 1971. *The Opening of the Canadian North 1870–1914.* Toronto: McClelland & Stewart.

Zaslow, Morris. 1988. *The Northward Expansion of Canada 1914–1967.* Toronto: McClelland & Stewart.

8

Modernism, Water, and Affluence: The Japanese Way in East Asia

Gavan McCormack

THE WAY OF AFFLUENCE

It is a commonplace that, for individuals and societies alike, powerful memories which are especially bitter and negative tend to persist longer than positive or happy ones; in the case of nation-states and their peoples, bitter memories may cloud perceptions of transformed historical contexts and block the process of necessary adaptation to change. Thus in late twentieth-century East and Southeast Asia the social memory of Japanese aggression and militarism of 50 or 60 years ago is kept alive, partly by the force of the horror it entailed and partly by the way in which Japan continues to equivocate about whether to celebrate or apologize for its actions then. In contemporary East and Southeast Asia, the phrase "the Japanese threat" is invariably interpreted in terms of the militarism of a half century ago, although few, if any, observers of Japanese society detect any signs of a revival of Japanese militarism.

This chapter argues, firstly, that the countries of the former Japanese "Co-Prosperity Sphere" are threatened today in quite specific ways that differ fundamentally from those of 50 or 60 years ago; however paradoxical it may seem, they are threatened, not by Japanese militarism, but by the Japanese model of affluence; secondly, that this model may be observed in particular in the way it impinges on attitudes and policies relating to water. While a significant paradigm shift is underway in "modern" attitudes to water in the advanced industrial countries, bureaucratic and corporate resistance to that shift is particularly strong in Japan, with the result that the outdated "modernism" retained at the heart of the Japanese political economy continues to be replicated throughout the region, at huge social and environmental cost.

The contemporary Japanese "way of affluence" is immensely powerful precisely because it is seen as "innocent" and as a far more attractive option than militarism, but it constitutes a threat precisely because it is neither sustainable nor reproducible, and therefore the longer the effort to sustain and reproduce it continues, the greater the ensuing damage and cost. Modernity, especially the patterns of consumer life by which it is commonly defined, often seems more accessible, more realistic, in its Japanese mode. Japanese capitalism as model and goal has long constituted a powerful transformative force, perhaps even from the time of the Japanese victory over Russia in the war of 1905, but certainly from the time of the Indochinese War of the 1960s and 1970s, when Japanese consumer goods began to displace American goods in the markets and on the roads of Southeast Asia, and from the 1980s when advertising and television began to refract images of Japanese urban consumer life throughout the region. The persuasiveness of the Japanese economic model was reinforced by the hyperbole, which during the 1980s surrounded its apparent "successes" ("Number One"), both in the Western literature and in much of the Japanese, all of which was refracted and even less critically analyzed in the East and Southeast Asian region. It was also reinforced by the active intervention of Japan, as promoter and financier, in the replication of that system. The Japanese voice in Asian institutions, from the Asian Development Bank to APEC or the Mekong River Commission, was also clearly the most powerful, and the tendency for Japanese bureaucrats and corporations to reproduce regionally (and also to a degree globally) the patterns of their home structure was inevitable. In the 1990s, the direction of growth, GDP expansion, and affluence has been toward a Japan-type goal. Because the roots of the system, both institutional and philosophical, are deep, change will not be easy.

It may be objected that Japan is merely one among many advanced capitalist countries. True, but it is the predominant capitalist country in the region—a colossal Mt. Fuji towering over its neighbors, accounting for nearly 70 percent of the economy of the vast region between Mongolia and New Zealand or 17 percent of the global economy, and while it shares the general characteristics of advanced capitalism, at the same time it embodies them in a form which is to some extent specific and culturally distinct. A substantial literature has been devoted to elucidating this "Japaneseness" in terms of organization of the economy, especially management theory, but relatively little to the questions this paper addresses. It is not a matter of a "Japanese" model antithetic to a "Western" one, but that much of the force of capitalism in East and Southeast Asia is refracted in Japanese cultural tones and shades which lend it a particular force.

At Rio the Japanese government formally committed itself to the principle of sustainability, but it soon made it clear that what it meant by that was the sustainability of growth, not the sustainability of society. The "way of affluence," apparently beyond ideology, is almost irresistibly seductive, and the message that "affluence" and sustainability might be irreconcilable contradicts central articles of contemporary faith and is therefore unpalatable. It has yet to be generally understood that the present bases of economic life in the country are themselves unsustainable, as well as being nonreproducible elsewhere. Generally speaking, the negative lessons of Japanese development over the past three or four decades are rarely considered and the "Way of Affluence" is pursued without reservation.

While China and Southeast Asia set their sights on replicating Japan's "success" and aspire to share its "affluence," the fact is that Japan offers also many of the elements of a counter-model: the deliberate abandonment of one of the world's most sustainable and environmentally benign agricultural traditions, combined with the pursuit of unsustainable levels of urbanization, consumption, waste, and pollution, a preference for private over public transport, and a pattern of consistent failure at regional development policy through the postwar decades under the series of "Zensò" or "Comprehensive National Development Plans" (McCormack 1996). The projection into the twenty-first century and on a regional basis throughout East and Southeast Asia of the Japanese way of mass consumption and mass waste, infinite manipulation of human desires in order to promote "growth," abandonment of the rural sector and concentration of the population in megalopolises, where the links between human and natural orders are continually eroded and the patterns of life are continually accelerated, is nothing short of a high road to catastrophe.

If all countries followed the Japanese path, "five or six planets would be needed to serve as 'sources' for the inputs and 'sinks' for the waste of economic progress" (Sachs 1995, 6). Here again the point may be expressed in both general and specific terms: global capitalism, sometimes described as triumphant in the wake of the Cold War, is actually nearing the end of its historic role. As the Japanese scholar, Furusawa Kòyù, notes, global GDP rose 33 times between 1900 and 1986 (from $0.6 trillion to $13.1 trillion in 1980s dollars), and if the 3 percent growth predicted by the IMF were to continue through the coming century, it would expand again by 2.7 times to the year 2024, 6 times to the year 2057, and 12 times to the year 2090, while global energy output would have to multiply by 2.2, 4.5, and 8.0 times, respectively (Furusawa 1996a). The absurdity implicit in such projections is plain, for where would the energy and materials come from to sustain such levels of productivity, and where would the wastes be disposed? The improbability that there can be any technological fix for this has profound implications. The message that capitalism, with its extraordinary capacity to feed expansion, can only be seen as part of the problem of the twenty-first century, not its solution, may seem bleak and heretical but must be considered very seriously, for it is hard to imagine that long-term sustainability can mean anything else but a goal of zero growth and zero waste.

Japan, the "success story" of East Asia, is embedded at the heart of this contradiction. Though all the region struggles to emulate and replicate what is seen as the Japanese "success," the fact is that such replication is impossible. Thus, for example, Japanese consumption patterns now call for the appropriation of two-fifths of world shrimp and nearly one-half of its tuna (one-quarter of all traded marine products) and also a large proportion of beef and other meats, grains, and increasingly fruits and vegetables. They entail the appropriation of the agricultural produce of a land area beyond its own borders two and a half times the size of that which lies within them, not to mention the oil, iron, coal, copper, lead, and other nonrenewable materials that are swallowed in the inexhaustible system of production, consumption, and waste. As the region, and the world, faces steady increases in population and small ones in production, an actual decline in availability of grain and fish, and the gradual run-down of reserves of fossil fuels, the continuing Japanese need to appropriate on this scale sharpens a fundamental contradiction between it and its neighbor countries. Far from it being a model to be emulated, it is the most serious threat to the creation of an equitable, prosperous, and sustainable future order. In the long term, it may not be possible to preserve existing Japanese privilege in terms of access to food and raw material resources without some mechanism of coercion.

In terms of justice and equality, the North (including, but by no means confined to Japan) cannot deny to the South what it insists on enjoying itself. The injustice of a world system in which the 20 percent of the world's people living in the rich industrialized north consume 80 percent of the world's energy and materials is clear.[1] The desirability of or right to development on the part of the "developing" world is not an issue, but the problem is not just one of "developing" the un- or underdeveloped South; it is deeply embedded in the "developed" (or "overdeveloped") North, of which Japan is both epitome and, in East and Southeast Asia, principal avatar.

WATER AND CIVILIZATION

In the age of Bill Gates and the information revolution, it is easy to forget that all civilizations depend ultimately on their solution to the basic problems of providing food, water, and shelter to their people and that the failure to achieve an ecological balance has been responsible for the failure of more than one great civilization. Water may be seen as the single most important factor in determining sustainability, and its role in the modernizing and industrializing process now appears in relief because of the considerable debates it has occasioned in recent years and the indications that a major shift is currently under way in paradigmatic thinking about it (McCormack 1997).

The aspiration to control the natural flow of water in the world's rivers is as old as civilization itself. Just as ancient civilization was born of the accommodation reached between the communities that once lived around the great rivers, the Tigris, Euphrates, Nile, Indus, Yangzi, and Yellow, so in a sense modern civilization, in its most concentrated American epitome best seen in Los Angeles and Las Vegas, was built upon the power harnessed from the Columbia and Colorado Rivers. The age of the super dams was ushered in by the Hoover Dam (221 meters high) that commenced operation in 1935 (McCormack 1997; McCully 1997). Where there had been only 50 dams constructed in the continental United States in the first three decades of the twentieth century, 1,000 were built between 1930 and 1980; after that, however, virtually none (Sumi 1996, 23). From the perspective of the late 1990s, the "age of dams" began in the 1930s, lasted 50 years, and then ended.

Despite the massive expenditure on dams and other flood control measures designed to "tame" the rivers, many existing dams are now seen to be unsafe, while flood damage rises across the country (Lippert 1996). When the head of the

U.S. Bureau of Reclamation, Daniel Beard, declared in 1994 that the era of large dam construction was over and insisted that it would be "a serious mistake for any region in the world to use what we did on the Colorado and Columbia Rivers as examples to be duplicated," reverberations quickly spread around the world.[2] Since it was the example of the damming of these rivers 60 years ago which seized the world's imagination and inaugurated the age of large dam construction on a global level, the *volte face* was very significant. The modern view of the river as a bundle of functions—of which power and irrigation were central—that may be exploited most efficiently by the appropriate mix of economic, engineering, and agricultural policies and structures is evidently coming to an end.

Early in the 1990s, the end of this modern paradigm was eloquently symbolized by the undertaking of works to demolish some dams, notably those on the Elhwa River in Washington State, in order to restore once bountiful fisheries (Sumi 1996, 53). Billions of dollars will be required in coming decades to redress some of the damages caused by the dams. As Beard noted, those problems were akin to those of nuclear power plants: "You get immediate benefits, but also long-term costs of a very great magnitude" (Beard 1996).

Like so much of the American culture of modernism, however, the secular faith in the dam spread with irresistible force over the rest of the world and only began to lose some of its force as the end of the century approached. Following the U.S. example, the phenomenon of damming the world's rivers began in earnest in the 1950s. The wave of postcolonialism and nationalism that swept much of Asia and Africa in the late 1950s was characterized by the combination of rejection of Western political hegemony with adoption, often in passionate measure, of a Western faith in nature-dominating technology, with the dam forming a centerpiece in that faith. As India's Nehru put it, dams were "the new temples of India, where I worship" (Pearce 1992, 3). From India, through Pakistan, Turkey, Iraq, Libya, Egypt, Ghana, Zaire, Zambia, to Brazil and Paraguay the same faith spread, while in the first and second worlds too its adherents remained strong. In retrospect, the Cold War years are more remarkable for this shared faith than for ideological differences.

More than 85 percent of the world's 38,000 large (more than 15 meters high) dams were built during the 35 years since 1960 (Postel 1996, 45). Grand projects, such as the Soviet-supported Aswan of the 1960s and many others were supported by the World Bank and Western financial, banking, and construction groups throughout the world. But while huge projects proceed, in places like Sarawak, Vietnam, and China, the debate and to some extent the reversals, on recent projects, most prominently Arun in Nepal and Narmada in India, suggest that the wave may have passed its peak and begun to recede. By 1995, around 15 percent of the runoff of the world's rivers had been appropriated, and "optimistic" projections called for that to be raised to two-thirds by the early twenty-first century (Pearce 1992, 134). The transformation of rivers from free-flowing phenomena of nature to elaborate plumbing systems has been characteristic of the late-century decades. Doubts about the human capacity, or right, to control and subjugate nature played little part in decision-making, although the ecological functions of rivers and aquatic systems were little understood. From the United States to the Soviet Union, and then across the world in a modern consensus that transcended any considerations of ideology, states poured massive subsidies into aquatic engineering works, persuaded that multiple benefits would flow from the appropriation of the power of the rivers.

In Japan, the first concrete dam was built in 1900. Large hydroelectric projects were undertaken during the 1930s, especially in Manchuria, but it was in the postwar period, after the arrival of the bulldozer, that the real rush began: 1,000 dams were built between 1956 and 1990; a further 570 are either under construction or at various planning stages; although 75 percent of the hydro potential of the ten main river systems is now tapped, the system is deeply entrenched and construction continues, moving ever further up into the tributary river systems deeper in the mountains. By the mid-1990s, monsoonal and mountainous Japan actually had built more dams than the continental United States (although much less in their capacity)[3] and was continuing to spend at the rate of around Y200 billion (roughly $2 billion) per year on new dam construction works (*The Nikkei Weekly,* 9 September 1996). At the same time, the rivers themselves were commonly concreted on their bottom and sides (the process known as *sanmenbari*) which supposedly made them safer and more subject to human control; the country's coastline was also concreted, its tidelands and some of its major lakes drained and reclaimed; and its forest watersheds eroded by multifaceted developments. Modernism was embraced with a passion, and the Shinto or Daoist elements in Japanese culture were steamrollered by a fusion of Confucian abhorrence for nature with positivism and modernism.

In key respects, China and Southeast Asia are following the Japanese path: sacrifice of agriculture to industry, of up-country and mountain regions to coastal and down-river regions, of health and environment to growth. In terms of energy, China's current (1996–2000) ninth Five-Year Plan is based on the assumption of GDP growth, which recorded the figure of 11.8 percent between 1990 and 1995, reducing gradually to average 8 percent in 2000, and then to a 7.5 percent average for the first decade of the new millennium (*Nihon keizai shimbun,* 16 October 1996). Power output will obviously have to grow substantially. Current capacity is less than 200 million kilowatts (hereafter KW, with 1,000 KW equal to one megawatt or MW), with 250 million KW needed by 2000 and 400 million KW, more than double current capacity, by 2010 and a further doubling, to 800 million KW, in the decade beyond that, a quadrupling process over two decades. The questions raised by such plans involve not only all Chinese, but the region, indeed the world.

To cope with this huge leap in demand, China is planning rapid expansion in thermal, nuclear, and hydro plants. Most of its coal is of the high sulfur-content variety that is a major cause of acid rain and associated ailments throughout the region, but because so much of it is available, it remains central to the national power generation. China's current annual emissions of sulfur dioxide are about 20 million tons (as against less than one million in Japan and Korea). Levels of asthma and other respiratory ailments and lung cancer are already high in the industrial areas, and acid rain erodes the region's forests, but industrial output is planned to double within the next 10 to 15 years.

Till the mid-1990s, China had few nuclear power facilities, and nuclear power accounted for an insignificant proportion of the national power grid, but that is set to change very quickly. The kind of increase envisaged in current plans to be realized—from 2.1 KW capacity in 1994 to 15 KW in 2010 and 30–40 KW in 2020—will involve transforming the East China coast into a kind of "Nuclear Power Station Ginza" to rival the concentration of similar facilities along the Japan sea coast in Fukui prefecture.

Hydropower, which currently accounts for about 20 percent of the national energy grid, is projected to rise to 30 percent by the end of the century, although initial construction and subsequent transmission costs exceed even those of the nuclear plants. The three great rivers originating in Chinese territory in the Tibetan plateau, the Yangzi, Mekong, and Salween, are all subject to plans with enormous implications for the ecology of East and Southeast Asia. Here we consider only the Yangzi and Mekong. In its plans for the development of these two great rivers, China is pushing its interventions in nature to new limits.

YANGZI: "THREE GORGES" DAM

The Yangzi flows through the heartland of China, with 390 million people living in its basin. The dam planned now for its middle reaches will alter the river in a way more profound than anything in Chinese history. The main wall of the projected "Three Gorges" Dam at Sandouping, just downstream from the Quting, Wu, and Xiling Gorges, will be 185 meters high and 2 kilometers long, holding back an immense 39 billion cubic meters of water as the river doubles in width, 632 kilometers of land is flooded, and the newly formed reservoir stretches back 650 kilometers to Chongqing, equal to the distance from London to Glasgow or Tokyo to Aomori. Power generation capacity will be 18 million KW. The dam will top the existing world's largest by 40 percent. It will inundate and submerge hundreds of cities, counties, towns, and villages and some of the most productive farmland in the country; it will force the relocation of around 1.5 million people (McCormack 1997; Qing 1994).

The project is truly pharaonic in scale. It will be the biggest work in China since the Great Wall and very likely the biggest civil engineering work in history. The project is marked by a spirit of techno-nationalism and hubris. It forms the centerpiece of the Ten-Year National Plan, and since 1989 debate on it has been closed and opponents silenced and prevented from publishing, although when the crucial vote was taken at the 1992 National People's Congress, a remarkable 32 percent either abstained or opposed it. For the Chinese leadership, this is unquestionably a project designed inter alia to "out-Hoover" the United States, to show the world in a dramatic way that it will be "No. 1," at least in its capacity to generate power, manipulate people, and dominate nature.

The arguments about all this are complex—technical, financial, environmental, and social; central is that of people since the uprooting of people will be of an unprecedented scale, and there is actually nowhere for the million and a half displaced people to go, but the consequences for various endangered species from the natural world will also be incalculable. Above all, critics protest the closure of debate and the manipulation of technical, financial, and environmental data, and insist that the scale of the project and its implications for future generations demand nothing less than full and free debate.

The U.S. Bureau of Reclamation and the Army Corps of Engineers, deeply involved in the early and planning stages of the project, withdrew technical cooperation in 1993, when it was announced that they were "into water resource management and environmental restoration, not large dam projects."

Critics also suggest that a collapse of the dam wall, whether caused by construction fault, earthquake, or enemy attack, would trigger an enormous cascade of water onto the cities of Wuhan and Changsha, such that the "scale of catastrophe and the number of deaths would defy the imagination" (Béja 1996). Huang Wanli, the veteran engineer who had worked in the 1930s on the Tennessee Valley project in the United States and whose opinion on the Yangzi project was sought by U.S. President Clinton in the 1990s, insisted that, even if the dam were to be built at huge financial, social, and environmental cost, sooner or later it would have to be blown up and destroyed in the interests of safety (Amano 1996).

The modernist vision to which the patriarchs of China were clinging on the cusp of the new century was a powerful blend of origins in early twentieth-century American technology and dam engineering, with strong influences from the Japanese fascist and Soviet Stalinist state-building and engineering phases of the 1930s and 1950s, topped off with the will to follow and surpass the Japanese as civil engineering superpower of the late twentieth century. Clearly, the modernist vision retained much of its pristine force in China. It is a phenomenon of which Wittfogel might feel justified in commenting "I told you so." The same rethinking of the basic principles of development and water engineering that was taking place throughout the late twentieth-century industrial world would no doubt come to China too, in due course, but it might come too late to save the Yangzi River.[4]

THE MEKONG

While the Three Gorges Dam has become relatively well known, the series of projects on the Mekong is in sum even greater in scale and presents problems that are in some senses more complex. The Yangzi flows only through Chinese territory, but the Mekong flows 4,200 kilometers from Tibet, through Yunnan, Burma, Thailand, Laos, Cambodia, and Vietnam. The river basin has an area of 795,000 square kilometers (more than twice the area of Japan). Millions of people depend for food, water, and transportation on the river system, which is host to one of the world's biologically richest aquatic communities. The development imperative is felt in this region just as it is in China, however, and projects for highway and railway construction, fiberoptic telecommunications networks, power and other infrastructure developments proliferated in the late 1990s, fed by the prospect of a newly-emerging dynamic economic bloc of 230 million people (Miyauchi 1996). The Mekong River Commission estimates a tripling of demand for energy in China's Yunnan province plus the Southeast Asian Mekong states by the year 2010.[5] Because of its still largely untapped potential, the Mekong is seen as the provider of cheap, clean energy to power the modern transformation of its watershed countries and to launch the whole region upon its "way of affluence."

Post-Indochina War, post-Cambodian peace settlement, and in the context of the triumph of ASEAN and the extraordinary transformation of China, development plans for the Mekong proliferate, and the engineers, bankers, and various development gurus of the virtually defunct dam industry of the United States, Europe, and Australia now converge on this region to apply and develop their skills.

The Mekong is commonly thought of as a Southeast Asian river, but it is also a Chinese river, and if the 14 dams already complete or under construction on the upper reaches of the river that lie within Chinese territory (in Yunnan province) were to proceed to completion, the total capacity (at 22.3 million KW) would be considerably greater than that of the Three Gorges (at around 18 million KW). Possible further developments with a total capacity of 40 million KW are under consideration, and from the fact that, according to *The Economist,* just one of the planned dams would produce a reservoir capable of holding the entire flow of the Upper Mekong for six months, the immensity of the scale of these plans may be guessed.

As for Southeast Asia, in the early 1990s a notion of "run-of-river" Mekong dams on the river's main stem was put forward; supposedly these would have little storage capacity and their generating output would be proportional to the flow at a given time (Rothert 1995, 20–21). The nine projects that were described as most promising had a total capacity of 14 million KW. Although described as "run-of-river," however, the structures envisioned were very substantial, involving walls of 35 meters, inundation of 1,000 square kilometers, displacement of more than 60,000 people, and the creation of reservoirs ranging from 75 to 200 kilometers in length (Rothert 1995, 23). The sites possess extraordinary natural and social diversity.[6]

Laos in particular is a poor and landlocked country of about four million people. It could be described as pursuing a hydropower path to nation-building, with the (rather ambiguous) prospect before it of becoming the Kuwait of Asia.[7] Around 35 percent of the Mekong's waters originate in Laos, and the estimated (theoretical) hydropower potential of that water alone is put at 13,000 MW. In the mid-1990s there were only three hydropower dams, generating 200 MW, but by agreements made in 1995 and 1996 with its neighbor countries, Laos is committed to providing between 1,500 and 2,000 MW annually to both Vietnam and Thailand by the year 2010. That will mean multiplying its existing capacity 20-fold, and memoranda of agreement have already been signed for the construction of 23 dams, many of them very large, within that period, with a further 30 (by some counts 50) dams on the drawing board or at various stages of planning. If all were built, the current 200 MW grid would increase by nearly 60-fold. Laos has few other sources of possible income, and the view of the river resources of Laos as like "sleeping beauties, waiting for their Prince Charming" is attributed to the Thai Foreign Minister, speaking in 1990 (Ryder 1996–1997, 16).

As modernizing Japan abandoned its upriver and rural hinterlands in Tohoku, Chùgoku Shikoku, and Kyushu, turning them into deserted realms for the old and useless in order to create a prosperous Kanto, Kansai, and Chubu, so China and Southeast Asia move likewise to abandon or sacrifice their uplands, most significantly in places like Laos, in order to bring an electrified, urban, convenient modern affluence to their downstream urban areas.

As for Vietnam, after 15 years of construction works and at a cost of about $1.5 billion, Vietnam completed in 1994 the construction of the massive Hoa Binh Dam, with 1,920 MW generating capacity. Covering a 200 square kilometer area, with a 770-kilometer shoreline and 9.5 billion cubic meter capacity, it became the largest in Southeast Asia (Houghton 1996). Despite the huge cost, the benefits were dubious. Even as it was completed, the dam's anticipated life had already been reduced by half to 50 years due to rapid erosion and siltation, while the benefits from irrigation and power generation were proving to be well below anticipation.[8] But Vietnam was already pursuing prefeasibility assessment studies of a dam of about twice the size to be known as Son La. This one, to cost an estimated $3 to $4 billion, would become the largest in Southeast Asia. Designed to produce around 4,000 MW of power, it would inundate another 440 square kilometers of forest and force the relocation of over 100,000 tribespeople (*Watershed* 1996).

As plans proceed for the sealing up of the Mekong upstream, both its main course and its tributaries, the consequences for the economy, society, and ecology of the river downstream are impossible to predict accurately. The ecology of the entire region is delicately balanced. As the forests are depleted and the flow of the waters interrupted, the rivers carry greater loads of silt, the fish catches decline—reports in many regions tell of a halving since the 1970s—and fishing people resort to use of finer mesh nets or even explosives, which can only worsen the long-term outlook. Floods come with greater frequency and violence, and the bed of Tonle Sap, the vast Cambodian lake formed each year by the reverse flow of the river waters in flood, rises as mud and silt accumulate, so that where the fishermen of one generation could stand up to their chests in water, their sons stand in knee-deep water. The Cambodian leader Sihanouk worries that only mud will be left for their children (Takabe 1997). Beyond that lies the prospect of the Tonle Sap becoming a Southeast Asian Aral Sea, a catastrophe of unimaginable proportions.

The potential for ecological havoc, and accompanying social and political unrest, is no less than that which is developing on the Yangzi, but it is complicated in the Mekong case by the number of independent states involved. From the point of view of the minority mountain- and forest-dwelling peoples who in effect have acted as guardians for humanity of an immense ecological treasure house, the prospect must in future be seen in terms of the prospect of developmental genocide as their livelihood and culture is placed under sustained attack.

SHIFTING PARADIGMS AND UNEVEN DEVELOPMENT

Ironically, even as the home of modern development, the United States is turning against the construction of large dams and finding economic benefits in demolishing some existing ones in order to restore natural aquatic ecosystems and thereby "to solve water quality, wildlife, and flooding problems at minimal cost and disruption," and the realization is beginning to grow (in the society if not yet in the bureaucracy) in Japan that the costs of dam development over the last 40 years have greatly exceeded the benefits and would continue to weigh upon generations yet unborn, the state bureaucracies of China and Southeast Asia continue to embrace outdated technologies, and the international financial and aid bureaucracies, especially those based in Tokyo, continue to encourage regional forms of development that would be impossible at home. As always, development proceeds most unevenly.

By reason of its economic scale and its experience and expertise in dam construction, Japan is involved in both the Chinese and Southeast Asian dam projects, and it plays the key role in the Mekong River Commission. Both the Three Gorges and the Mekong projects featured prominently in the list of Global Super Projects drawn up by a high-level committee headed by Okita Saburò and promoted by the Japanese government in the late 1980s, and in the mid-1990s Japanese support in finance, technical advice and consultation, or actual construction was eagerly anticipated by neighbor countries. At a deeper level, it was the desire to achieve the impossible, even the absurd—a Japanese-style standard of consumer life—that was driving the whole transformative process.

The prospect at the moment is that the change in thinking about rivers and water, part of a civilizational reappraisal of value and modernity, will be too late for East and Southeast Asia. In this respect, Japanese ambivalence is highly significant: the Japanese commitment to modernism has been unequivocal and passionate, and the civil engineering and construction sectors have been central to the Japanese political economy, the functional equivalent to the defense-industry complex in the United States. The Japan that was proclaimed "No. 1" less than 15 years ago now faces immense problems. The fiscal crisis is clear, and I have discussed it elsewhere. Even more serious may be the indications that Japan has lost its faith and has little to offer in terms of ideas and vision for the future of humanity. The prescriptions that are currently debated—deregulation and a financial "big bang"—scarcely begin to address the underlying structural issues of value and direction. I have written enough elsewhere about the tragic costs of the Japanese developmental model, with its population concentrated in the megalopolitan Pacific belt, linked by expressways and super express trains, while its mountain or inland hinterland is semideserted, either left to the old, used for dumping industrial waste or constructing dams and nuclear power plants, or for constructing golf and other leisure installations. An actual identity of hyperwork, consumption, and waste in a world of concrete is combined with a virtual reality in which greenness, tradition, nature, and harmony are preserved intact.

So far as energy is concerned, there is no question but that China and Southeast Asia face a large energy problem. What is questionable, however, is the assumption that the problem can only be met either by the huge expansion of conventional large utility (thermal, hydro) or nuclear power grids. There are two reasons for this: the inefficiencies of the existing grid, and the increasingly realistic possibility of alternatives to it. The inefficiencies of the existing power generation and transmission system are immense, and the scope for savings by application of existing technologies has yet to be thoroughly explored.

Not only in China and Southeast Asia, but globally, dependence on oil, coal, nuclear, and hydropower energy will have to give way to nonpolluting and renewable energy sources during the twenty-first century, and a reversal of the current trends towards megalopolitan concentration of the earth's people will be a necessary precondition to achieve this (Murota 1996). Fortunately, estimates of the potential for renewable energy generation have recently been drastically revised so that, for example, the wind from just a few states in the western United States is now estimated to be sufficient to provide the electricity needs of the entire country, while the conversion of even a modest 3 percent of China's extensive wind power potential (estimated at 1,600 million KW) to electricity would be sufficient to meet 25 percent of its current national requirements (Byrne, Bo, and Xiuguo 1996). Geothermal reserves and the prospects for solar (photovoltaic) technology in China are also said to be excellent.

A number of radical alternative proposals have been formulated by democratic and locally based environmental and social movement spokespersons. The influential Southeast Asian movement-based *Watershed* (TERRA) proposes a decentralization of electricity supply, involving rapid expansion of micro-hydro, solar photovoltaic, biomass gasification, and small-scale windmills "controlled and maintained through community-based electricity supply management systems" as elements of a solution that is already available and which would have little or no impact on the natural environment of local communities (*Watershed* Editorial, 1996–1997).

Likewise in China, Dai Qing is part of a widespread and deep-rooted regional movement pursuing not only the best and most appropriate technology but a radically different vision of development and affluence. Instead of the massive Three Gorges project, she suggests a decentralized pattern of smaller dams, combined with large-scale reforestation works, on the upper reaches of the Yangzi River, and general repair work on the banks and dykes of the river for the clearing of silt, etc., and the exploration of alternatives for efficient and cheap transport between the coast and the interior. She also points out that resources badly needed for other projects, in education, transportation, and communications will be drained away by this one mammoth and risky project, and she notes that one-third of dams built in China since the 1950s are unsafe, pointing to incidents of collapse in 1975 and 1993.[9]

The German Wuppertal Institute has begun the attempt to spell out the concrete meaning of commitment to sustainability in practice, and the Sustainable Europe conference held in Brussels in November 1995 was a step toward erecting some signposts toward that goal (Furusawa 1996b). Wuppertal's position is that "no more of a renewable resource should be utilized than can regenerate in the same period" and that "only that amount of materials should be released into the environment that can be absorbed there." At the same time, the use of energy and throughput materials would have to be cut to a low-risk level, and an equitable balance in access to nonrenewable energy and raw material resources, which would involve a massive reduction in the flow of such materials to the industrial North, would have to be achieved (Wuppertal Institute 1995). No industrial or developing country would currently meet such standards, and some, Japan among them, face a major crisis of adjustment when they come to appreciate what is involved. For Germany, Wuppertal envisages the stabilization at existing levels, currently 11.3 percent of the national land, of the settlement and transport infrastructure—in other words, an immediate and total halt to encroachments on the land in the name of urban or transport development; a slashing of carbon dioxide and of nitrogen gas emissions by 80–90 percent over the next several decades; a total phasing out of nuclear power; the switch from chemicalized to organic agriculture; the reduction of fossil fuel use by 80–90 percent by the early twenty-first century; and the rapid expansion of use of renewable energy sources, including wind and solar power.

These are the necessary steps to achieve the commitments entered upon at Rio (and subsequent conferences), but they require an immense transformation in existing society. No such detailed prescription has yet been formulated for Japan, but it is unimaginable that the prescription there could be any less drastic. Though Japan is widely regarded as a model of energy efficiency because of the huge improvements achieved under the impact of the oil shocks in the 1970s, the fact is that the real gains it made then were quickly swallowed up in the increased consumption of energy and nonrenewable raw materials stemming from the growth of the economy as a whole. The sense of urgency and need for structural reform was sapped also by the high yen and the cheaper resulting energy prices from the late 1980s. In terms of the comparison with the United States, Japan's record still looks good (Yukako 1995). Its energy efficiency per unit of GDP produced was (in 1992) more than double that of the United States (and at least ten times that of China) (*Nihon keizai shimbun,* 23 October 1996). However, compared to other industrial countries its performance gradually weakened, and energy consumption between 1986 and 1995 grew at about 150 percent (*Asahi shimbun* Editorial 2 December 1996).

Despite the official commitment to hold CO_2 emissions to 1990 levels by the end of century, Japanese emissions were steadily rising through the first half of the decade, and the bureaucratic prescription to cope with the energy problem remained frozen in the mode set during the oil crisis of the early 1970s, with priority given to securing guaranteed access to energy raw materials and the quest for autonomy from global market forces through increasing reliance on nuclear power and development within Japan of the full nuclear cycle. The nuclear contribution to the national power grid rose by 81 percent during the decade up to 1995 (to 30 percent of total energy, with 53 reactors), and the bureaucratic plan focused on increasing that to 45 million KW by 2000, 75 million by 2010, and 100 million KW by 2030. So deep is this commitment that around 50 percent of the global environment protection budget allocation was in the mid-1990s being appropriated by the nuclear sector (Ueno 1996, 149). By a similarly devious bureaucratic argument that Japanese CO_2 emissions would be reduced, and thereby global warming reduced, by the construction of more roads (allowing smoother traffic flow), the huge sum of 8.6 trillion yen (around $78 billion in 1996) was being allocated to highway construction (Fujita 1996, 149).

In per capita terms in the early 1990s Japan was consuming energy and raw materials, while producing CO_2 and other wastes, at rates far greater than would be its share under any (theoretical) global ecospace entitlement; producing annually 8.5 tons per capita of carbon dioxide to the European 7.3, both well below the U.S. figure of almost 20, as against a global ecospace prescription of 1.7 tons, with similarly excessive consumption of iron, aluminum, cement, and timber, all well in excess of the European figure and around ten times above its global "ration."[10] In other words, the efficiency gains of Japan and the OECD countries in the 1970s and 1980s are relatively trivial in the context of what is necessary. The industrial countries in general, including Japan, must "dematerialize" their economies (or reduce the physical flow from the ecosphere into the economy) by a factor of about ten over coming decades in order to achieve both a more equitable global access to world resources and to restabilize the ecosphere (Hinterberger, Luks, and Schmidt-Bleek 1995, 15).

FROM "MODERNITY" TO SUSTAINABILITY

The model of development that is being pursued with passion by the states of East and Southeast Asia is the current Japanese model of mass production, consumption, and waste, based on unsustainable levels of energy and materials exhaustion and leading to the sort of alienated emptiness that characterizes late twentieth-century affluent Japan. The hundred-year system of mobilization of people and resources to achieve economic growth, the national faith of GNPism that Japan has pursued and China and Southeast Asia adopted, is no longer viable.

While China and Southeast Asia consciously or unconsciously set their sights on replicating the Japanese economic development "miracle," the immense distortions and unnecessary costs paid to accomplish it are either little appreciated or are seen as part of the necessary price to be paid to achieve national greatness. The same techno-optimism to the point of hubris, insistence that grand nature-remolding engineering schemes constitute the core of national development, and disdain for the critical or dissenting views of local communities with different priorities that has characterized the Japanese trajectory over recent decades today may be seen in the postures of neighboring political leaders such as Li Peng and Mahathir Mohammed, as in others elsewhere such as Libya's Gadaffi or in prominent Asian and African statesmen of the postwar period such as Nehru, Nasser, Nkrumah, Kaunda, or Suharto. While the elites of China and Southeast Asia often claim to be rejecting Western and affirming Asian values, at a fundamental level they remain enthralled by the images of power and subjection of nature conveyed by the mighty symbol of the Hoover Dam, while people like Dan Beard are seen as mad and troublesome Western Daoists.

While in China and Southeast Asia the long-past threat of Japanese militarism is regularly, often almost ritually, denounced, the very different, contemporary threat of Japanese late-capitalism is largely ignored. Instead, the uncritical aspiration to reproduce the Japanese "success" story remains strong. In Japan itself, politicians, intellectuals, and the people generally have scarcely begun to address the sort of international contribution that would be required to meet the country's international responsibilities. Until the sort of prescription outlined by the Wuppertal Institute becomes the actual agenda for the industrialized countries, the critique of the path as chosen by the developing world will inevitably ring hollow, and an alternative to the massive hydro (and nuclear) power projects currently being pursued by the state bureaucracies of the region will be dismissed as hypocritical posturing. The Yangzi and Mekong issues are therefore at the same time both local and global, and they are issues that concern not only the regional economy and environment, but also the direction of civilization as a whole.

NOTES

1. Actually the UNDP estimates that the share of the world's 20 percent of "wealthy" peoples who live in the "North" increased from 70.2 percent to 82.7 percent in the years from 1960 to 1989 (Furusawa 1995, 33).

2. See Beard statement reproduced in *Watershed* 1.1 (July 1995): 6–9; also his speech to the Federation of Japanese Bar Associations in Okuma Takashi, Amano Reiko, Hobo Takashi and Daniel Beard, *Nihon no damu o kangaeru* ("Thinking About Japan's Dams"), Iwanami bukkuretto, No. 375, 1995, 37–55; also his regular contributions to the International Rivers' Network's *World Rivers Review* (Berkeley, CA).

3. With a land ratio of 1:25 between Japan and the United States, Japan had 2,576 dams (structures marked by a wall of more than 15 meters in height) as against 2,941 in the United States, but with respective capacity in million tons of water at 20.4 billion in Japan versus 433.4 billion in the United States (see Igarashi and Ogawa 1997, 114–15).

4. My personal, albeit superficial, impression from three days spent sailing down the Yangzi in February 1997 is that the river even undammed is in severe crisis: I saw no sign of anyone fishing or of any fish industry based on the river, which carries an immense amount of garbage.

5. Mekong Secretariat, "Actual and Projected Electricity Demand, 1986–2020," table reproduced in Rothert (1995, 17).

6. This author traveled down the Mekong in Laos between Pak Beng and Vientiane in February 1997.

7. Quoted from *The Nation* (Bangkok), 4 March 1994, in Usher (1996).

8. Houghton gives the figure of 62 million tons of sediment flowing every year into the Dam (Houghton 1996, 34–35).

9. Human Rights Watch Asia estimates that the 1975 collapse of the barrages of Banquiao and Shuimanquiao, on the Huai River, caused the deaths of around 240,000 people (quoted in Béja 1996).

10. See table in Furusawa (1996b, 11). The Japanese CO_2 figure given here, however, dates from the late 1980s, and by the mid nineties had already increased to 9.3 tons per capita (personal communication, Furusawa Kòyù, 21 January 1997).

REFERENCES

Amano, Reiko. 1996. Interview with Huang Wanli. *Shùkan kinyòbi* (29 November): 50–51.

Asahi shimbun Editorial: Chikyùjin no seiki e. 1996. *Asahi shimbun* (2 December).

Beard, Daniel P. 1996. Hard Lessons from the U.S. Dam-Building Era. *World Rivers Review* 11.4 (September): 6–7.

Béja, Jean-Philippe. 1996. La mégalomonie en action pour maitriser le Yangtse. *Le Monde Diplomatique* (Juin).

Byrne, John, Bo Shen, and Xiuguo Li. 1996. The Challenge of Sustainability: Balancing China's Energy and Environment Goals. *Energy Policy* 24(5): 455–61.

Fujita, Toshio. 1996. Shimin ni yoru Nihon kankyò hòkoku. *Sekai* (November): 146–55.

Furusawa, Kòyù. 1995. *Chikyù bunmei bijon* ("Vision for a Global Civilization"). Tokyo: NHK Books.

Furusawa, Kòyù. 1996a. WTO taisei no daha ga kadai ("Breaking With the WTO System"). *Shùkan kinyòbi* (13 December): 28–30.

Furusawa, Kòyù. 1996b. "Chikyùteki kòsei" o Nihon shakai no shishin ni ("Global Justice as Guiding Principle for Japan"). *Shùkan Kinyòbi* (2 August): 9–11.

Hinterberger, Friedrich, Fred Luks, and Friedrich Schmidt-Bleek. 1995. What is "Natural Capital"? *Wuppertal Papers* 29 (March).

Houghton, Georgina. 1996. Vietnam's Hoa Binh Dam: Counting the Costs. *Watershed* 2.1 (July–October): 26–37.

Igarashi, Takayoshi, and Ogawa Akira. 1997. *Kòkyò jigyò dò suru ka* ("What Is to Be Done About Public Works?"). Tokyo: Iwanami shinsho.

Lippert, Suzanne. 1996. Time Enough for Change. *World Rivers Review* 11.3 (July): 4–5.

McCormack, Gavan. 1996. *The Emptiness of Japanese Affluence.* New York: Sharpe.

McCormack, Gavan. 1997. Food, Water, Power, People: Dams and Affluence in Late 20th Century East and Southeast Asia. *Kyoto Journal* 34 (April): 5–28.

McCully, Patrick. 1997. *Dammed Rivers.* London: Zed Press.

Miyauchi, Teiichi. 1996. Mekong Development Starts to Flow. *The Nikkei Weekly* (21 October).

Murota,Yasuhiko. 1996. Shin jidai no enerugii seisaku o tsukareru ka ("Prospects for a New Kind of Energy Policy"). *Shùkan Kinyòbi* (16 February).

Nihon keizai shimbun. 1996. Chùgoku no enerugi kankyò ("China's Energy Environment"), part 2, (16 October).

Nihon keizai shimbun. 1996. Chùgoku no enerugi kankyò ("China's Energy Environment"), part 8, (23 October).

Nikkei Weekly, The. 1996. 9 September.

Pearce, Fred. 1992. *The Dammed: Rivers, Dams, and the Coming World Water Crisis.* London: Bodley Head.

Postel, Sandra. 1996. Forging a Sustainable Water Strategy. In *State of the World 1996,* ed. Lester R. Brown et al., 40–59. London: Earthscan & Worldwatch Institute.

Qing, Dai. 1994. *Yangtze! Yangtze!* London and Toronto: Earthscan.

Rothert, Steve. 1995. Lessons Unlearned: Damming the Mekong River. International Rivers Network. Working Paper 6, October. Berkeley, CA.

Ryder, Gráinne. 1996–1997. The Rise and Fall of EGAT: From Monopoly to Marketplace. *Watershed* 2.2 (November–February 1997): 13–23.

Sachs, Wolfgang. 1995. *Global Ecology.* London: Zed Press.

Sumi, Kazuo. 1996. Amerika ni okeru damu kensetsu no rekishi ("History of Dam Construction in the U.S."). In *Amerika ha naze damu kaihatsu o yameta no ka* ("Why Did the U.S. Stop Dam Development?"), ed. Kòkyò jigyò chekku kikò o jitsugen suru giin no kai ("Diet Members League for Establishment of Controls over Public Works"), 22–54. Tokyo: Tsukiji shokan.

Takabe, Shin-ichi. 1997. Mekon monogatari (Mekong Story), part 2. *Yomiuri shimbun* (29 January).

Ueno, Hideo. 1996. Shimin ni yoru Nihon kankyò hòkoku. *Sekai* (November): 146–55.

Usher, Ann Danaiya. 1996. Damning the Theun River. *The Ecologist* 26.3 (May/June): 85–92.

Watershed 2.1 (July–October 1996): 5–6.

Watershed Editorial: The Politics of Power, 1996–1997. *Watershed* 2.2 (November–February): 2–3.

Wuppertal Institute for Climate, Environment, Energy. 1995. *Sustainable Germany: A Contribution to Sustainable Global Development.* English resume authorized by Wuppertal Institute on behalf of Bund and Misereor. Wuppertal, Germany: Wuppertal Institute.

Yukako, Fukasaku. 1995. Energy and Environment Policy Integration: The Case of Energy Conservation Policies and Technologies in Japan. *Energy Policy* 23(12): 1063–71.

Wastelands in Transition: Forms and Concepts of Waste in Hungary Since 1948

Zsuzsa Gille

WESTERN IMAGES OF SOCIALISM

Basic understandings of former socialist societies have long included the topic of waste. Most observers treated waste as a characteristic by-product of the low productivity and immense "inefficiency" of centrally planned economies, but a few others (including East European fiction writers and filmmakers) went beyond that and, for the purposes of social critique, designated waste as a metaphoric encapsulation of the essence of these societies. Yet another group emphasized the positive products of socialism (massive industrial and infrastructural investments, and the significant increase of living standards for large segments of the population, and so forth) and minimized the occurrence and amount of waste.[1]

Now, after the collapse of state socialism in Eastern Europe, its achievements are downplayed again, while destruction is stressed. A representation of socialism as a landscape littered with unwanted, low quality goods, garbage and dirt is juxtaposed with the images of the naturalness and purity of capitalism. Here the metaphor of waste plays a new ideological role. Waste reflects the irrationality of socialist economies, an irrationality understood as an absence of Western economic and technical rationality. Thus the impression is created that with Westernization, that is, with the rationalization of the economy, waste will be reduced, and thus the transition to capitalism will automatically benefit the environment. As the *Christian Science Monitor* said, "Many of the worst polluters . . . are being gradually shut down. More will certainly follow as obsolete and uncompetitive state-run enterprises in Eastern Europe continue to collapse as market economies take over" (Steichen 1991). Meanwhile *The Wall Street Journal* argued explicitly that "[n]ow all this [environmental pollution] is being swept away by democracy and economic rationality" (Solomon 1990).[2]

Nevertheless, waste in the literal sense, especially ongoing industrial waste, receives much less attention in these laudatory scenarios of the transition to market and democracy. While the catastrophic environmental records of socialist societies have been widely noted, waste-related hazards have been largely underrepresented even in the most recent literature on socialist environmental destruction, as well as in the targets of Western environmental aid (Schreiber 1991; Simons 1993). This ignorance is partially rooted in the myth that Eastern Europe's environmental problems are solely the "legacy of communism" (Hughes 1990), that is the legacy of the extreme wastefulness and squandering of natural resources by centrally planned economies. Such an expectation regarding waste is justified inasmuch as planned economies were characterized by extreme wastefulness and produced much higher amounts of waste (including hazardous waste) per GDP than did market economies (Table 9.1) (Castoriadis 1978; Goldman 1972; Filtzer 1986, 1992; Kornai 1980; Manser 1993; Nove 1980; Reiniger 1991; Simai 1990; Szlávik 1991; Ticktin 1976, 1992). But does that mean that they were more wasteful than Western countries? And more importantly, does that mean that implementing capitalism will end wastefulness?

Without denying the tremendous devastation of nature in former socialist countries, I would like to propose "No" as an answer to these questions. I wish to make two arguments in this paper. One, despite conventional wisdom, socialist societies were quite waste-conscious and some of their institutions to reuse waste were remarkable, even if mostly failed, social experiments that are now not only done away with but are forced into oblivion. Two, capitalism and existing socialism do not simply represent the good and the bad sides of the clean/dirty or the efficient/inefficient dichotomies, and therefore what awaits post-socialist societies is just a different form and different concept of waste rather than cleanup.

In the interest of a balanced view, I will start by explaining the systemic forms of waste in core capitalism and in semiperipheral state socialism. Then I proceed to discuss the conceptualizations of waste and the consequences of a politics of waste under state socialism and in what is now called the "transition to market and democracy."

THE MATERIAL FORMS OF WASTE IN CORE CAPITALISM AND THE FORMER STATE SOCIALIST SEMIPERIPHERY

In general there are two forms of waste: productive waste resulting from production inefficiencies, and allocative waste resulting from allocative inefficiencies. In Western core capitalism, allocative waste is generated by the wasteful post-facto adjustment of the market mechanism that was widely analyzed by Marx and Marxist economists. Allocative waste refers to already produced, functioning goods that cannot be sold or to goods that are hardly used but are discarded due to planned obsolescence (Baran and Sweezy 1966; Packard 1960; Toffler 1970). This form of waste is a systemic feature of capitalism.

Productive waste is due to material inefficiency, what Raymond Murphy (1994) termed the deviation from the ideal engine—that is, producing with more than the minimum of by-products. In theory, there is nothing in the profit motive that makes material inefficiency necessary, but there is also nothing in it to prevent

Table 9.1
Annual Nonmunicipal Waste Production and Its Ratio Per GDP and Per Capita (1000 Tons)

Country	Year	Waste	Waste/GDP	Waste/cap.
Canada	1990	1,079,390	1.90	40.55
USA	1990	9,880,710	1.81	39.33
Japan	1985	312,251	0.23	2.59
Austria	1990	33,852	0.21	4.39
Belgium	1988	30,989	0.20	3.11
Denmark	1985	3,836	0.34	0.75
Finland	1990	55,910	0.41	11.21
France	1990	559,800	0.47	9.92
West Germany	1990	96,736	0.06*	23.74
Greece	1990	15,974	0.24	1.58
Iceland	1990	135	0.02	0.53
Ireland	1984	26,500	1.49	7.51
Italy	1991	34,710	0.03	0.60
Luxembourg	1990	1,300	0.14	3.40
Netherlands	1990	28,819	0.10	1.93
Norway	1990	29,000	0.27	6.84
Portugal	1990	1,044	0.02	0.11
Spain	1990	195,902	0.40	5.03
Sweden	1990	66,475	0.29	7.77
Swizterland	1990	1,000	0.00	0.15
UK	1990	271,400	0.28	4.73
Czechoslovakia	1987	630,150	12. 08	40.47
Hungary	1989	107,493	3.69	10.13
Poland	1990	143,861	2.31	3.77
East Germany	1990	34,064	0.24*	2.10

* The GDP values for East and West Germany are based on UN estimates that were the same for current and constant prices (World Development Report 1990).

Source: My calculations from OECD Environmental Data (1993).

production with the higher than necessary waste ratios and with the more toxic by-products. Murphy (1994) argues that excluding waste costs (costs arising from treating or disposing of wastes or cleaning up waste-induced pollution) from other expenses of production makes this practice more the rule than the exception, leading to what he calls the "survival of the filthiest."

In socialism, we also find both allocative and productive wastes. The allocative form of waste here resulted from the tightness of the plans and from the strict sanctions involved in not fulfilling the plans. Companies had to hide their capacities and supplies and were interested in hoarding raw materials, even those that were not used by them directly but which they hoped to exchange for other inputs they did need. It was enough for one item to be missing to paralyze production even where all remaining inputs were otherwise available. These available inputs, which Hungarian economist János Kornai (1980) called slack, later themselves turned into wastes if unused—and they often were. Another type of allocative waste was defective products manufactured with substituted inputs, which thus often could not fulfill the needs for which they were designed. Due to shortages, these defective products were kept in the production process and ended up producing rejects themselves. Donald Filtzer (1992) provides a detailed description of this spiraling reproduction of waste in Soviet-type economies.

A Hungarian example comes from the May 8, 1951 issue of *Szabad Nép*.[3] A letter from the party secretary of the Ganz factory complains that "there are 30–40,000 kilos of high-alloy steel lying around on the yard, outside the warehouse in the open air, and when it rains, there are 10–15 cm. deep puddles under it. . . . [T]he rust eats away the etching" so no one can tell for sure what material it is. If its weight seems to fit, they "baptize the material" or throw it on the pile of unknown goods. To make up for the missing amount, those taking inventory cut off that much from another bar and paint the number of the missing material on it. Since it is unlikely to be exactly the same quality as the missing one was, the cut-off piece either becomes "unfit" for the technology it is needed for and/or ruins the machines. As a result, good quality materials become useless and then are sold at a discount to other plants. This form of waste is a systemic feature of state socialism, and since it results from tight plans and allocative inefficiency, let's call it "allocative waste."

The productive form of waste is the better known form of socialist waste. Waste as material inefficiency refers to the overconsumption of raw materials, the production of goods with a much higher material intensity than necessitated by technical prescriptions. Unlike in capitalism, here there was an explicit reward for both workers and management to produce with high material intensity because quotas were usually given in use value, mostly in weight, with little specification for quality. For example, they tended to manufacture heavier products than necessary, if producing fewer individual articles with more weight meant an easier fulfillment of the plan. A common practice among metallurgical companies was to fabricate large metal sheets rather than the sizes their buyers needed. Thus in plants where these sheets were cut into smaller pieces for pipes, tools and machines, the remainder constituted a much larger amount of scrap than necessary.

An article in the May 6, 1951 issue of *Szabad Nép* illustrates what such a motivation meant for customers of consumer goods, trying to buy 2–3 liter pots. In the basement of the store the author describes:

Pots and pans of 1020 liters are towering still from 1944! But the factories—despite all the requests, begging and notices—keep delivering the pots with large cubic capacity. . . . Why can't the customer get the pot s/he needs? Because the factories (Lampart, the enamel-ware plants of Köbánya, Budafok, and Bonyhád) fulfill their plan quotas measured in tons, ignoring the needs of the consumers, they only care about how they could fulfill their plans easier. . . . And the warehouses of Vasért (the company running hardware stores) in Vadász utca and Üllöi út provide the same experience, where a whole range of important although smaller screws are not available, while the warehouses have an abundance of large screws. The Screw Mill, the Hungarian Wagon and Machine Works fulfill their plans by weight: they deliver the screws on order but they lag behind in the production of the important small screws.

Several authors had brought attention to the outsize material intensity of planned production. A Hungarian electric engine weighs 15 tons more than an Austrian one with the same capacity, an agricultural trailer weighs 800 kilos more than its Austrian version, a harrow 140 kilos more, a seed-grain drill 210 kilos more, a quadruple-furrow plough 450 kilos more, and an oil-cooled converter 100 kilos more (Juhász 1981). Similar examples are available from other socialist countries.

We can also observe high material intensity on the macrolevel (Table 9.2). Gomulka and Rostowski (1988) analyze the ratio of intermediate inputs to gross value added for four COMECON countries and OECD countries and find that the average material intensity of socialist countries exceeds a corresponding capitalist measure by 26 percent in industry, 26 percent in manufacturing only, and by 43 percent in the economy as a whole; but the same percentages jump to 33 percent, 43 percent, and 52 percent, respectively, if energy prices are adjusted upward to the level of Hungary's and Yugoslavia's. The difference in energy intensity appears even higher according to the same authors. While the COMECON (for seven countries) average ratio of gross value added per unit of primary energy consumed in industry is 762 in 1965, the same OECD index (for 15 countries) is 1466, implying that the former use practically twice as much energy to generate the same final output as the latter. Moroney (1990, 212) similarly found that COMECON countries "as a group consume, on the average, approximately twice as much energy per unit of capital and per unit of real GDP as the economies of Western Europe."

Besides allocative and material wastes, however, in state socialism, and especially in semiperipheral East European countries with an economic dependence on the West and a political dependence on the Soviet Union, a third form of waste also existed, which originated in an unequal international division of labor. These countries were often forced by COMECON commitments and found it profitable to take over from the West the production of goods whose manufacturing entailed a large waste-per-final-product ratio, especially those that implied large amounts of toxic waste. In one case I studied, a Hungarian chemical

Table 9.2
Selected Material Intensities

Country	Steel use/GDP (units)	Cement use/GDP (units)
Hungary	100	100
Austria	46	88
Belgium	44	46
Denmark	36	38
France	47	50
Japan	72	no data
Great Britain	69	no data
West Germany	60	45
USA	68	no data

Source: Juhász (1981)

firm in Budapest undertook the production of tetrachlorobenzene (TCB), a herbicide intermediary, in 1968. Western producers were already trying to get rid of this nasty production process, which generated as much toxic by-product as final product. In a deal with Austria, in return for TCB, the Hungarian firm received much needed agricultural chemicals and hard currency. This was the first cooperation between a capitalist and socialist chemical firm in Hungary, itself a trailblazer in East–West economic cooperation.

Was the wastefulness of socialist countries simply due to their relatively low level of development? My statistical analysis suggests that it was not. The differences in waste/GDP data between capitalist countries and socialist countries in Table 9.1 are not attributable to the level of development measured by a GDP/capita index but rather to the mode of production. A regression analysis shows the former relation not significant, while the latter significant ($F = 0.007$), with the mode of production explaining about 60 percent of the change in the amount of waste per GDP.[4] This table, however, also alerts us to the fact that if by wastefulness we mean high waste per capita indexes, rather than high waste per GDP ones as is usually done, some capitalist countries, notably the United States and Canada, should be considered even more wasteful than most socialist countries. Taken together these data indicate the necessity of analyzing the different forms in which waste is produced in core capitalism and the state socialist periphery rather than assuming that the former is efficient and thrifty while the latter is inefficient and wasteful.

THE POLITICIZATION OF WASTE PRODUCTION AND REUSE UNDER EARLY STATE SOCIALISM (1948 TO THE MID-1970s)

State socialism's wastefulness might suggest that while it lasted there was a discursive silence on waste; the leaders wanted to cover up such squandering. In fact, this was my hypothesis. Yet what I found was that from the very beginning waste was a salient economic category that was subject to planning, and that tremendous political effort went into organizing the reuse of wastes produced by industry. In what follows I describe this politics of waste and offer some explanations before I move on to discuss its consequences.

Just as the material production of waste differed in core capitalism and state socialism, so did the conceptualization of waste. Already the ideology of the party marks a significant deviation from Western waste ideology, whose slogan could be "just throw it away." The designers of socialism projected a "New Man" and society, in which capitalist squandering ends, in which sparing materials is finally a worthwhile effort, since what is squandered is everyone's loss. The party and those who defined its morality often identified socialism as the "thriftiest of all societies" (Ember 1952), as "ending the outdated view that worships new materials" (Dömötör 1980; Jócsik 1977), and talked about "socialist man whose splendid feature is thriftiness" (Ember 1952).

But talk was not all. Numerous institutions were established to deal with waste. A key central organ was the Office for Saving Materials, founded as early in 1948 with the aims of searching for material resources, researching the possibility of raw material substitution, and reusing wastes. However, another "waste institution," the Secondary Raw Material and Waste Reuse Company (MÉH), which started its operation simultaneously with the first Five-Year Plan in 1951, had come to play a more significant role in individual lives. It was obliged to buy waste from all state enterprises, and its hands were also tied in terms of its customers. MÉH bought and sold metal, paper, textile, batteries, leather, wild berries, seeds and nuts, as well as animal bones and human hair. It also paid individuals for domestic wastes.

Legislation also indicated that waste was a significant economic category. In 1950, the National Planning Bureau issued a decree obliging production units to report the amount of waste in their possession, and in 1951 it established waste quotas and required enterprises to deliver their wastes to designated companies. The same year 12 more central regulations on the storage, delivery, and prices of waste materials were issued. Laws prescribed which wastes were to be delivered to which company, how to calculate the price of wastes, what to do with so-far unregulated waste materials, and how much material reward could be given to those who collected wastes beyond the planned amount.

The Hungarian bookkeeping system, developed during these years, also operated with the notion that waste is a useful material, which it tried to inculcate in economic actors. The account called "waste materials" was a material inventory type account, and wastes "regained," that is, taken back into warehouses, were to be represented on the "resource" side of the material balances as "other increase."

Numerous campaigns were organized either by MÉH or by various party organs that aimed at collecting layaway wastes and/or reusing them in factories, in agricultural cooperatives, in schools, and in districts of cities and villages. Metal-collecting weeks were organized; brigades dedicating themselves to waste reduction and reuse, waste-collecting stewards, youth and female troops mushroomed and busied themselves mostly beyond work hours. The next quote from an article about the campaign in the Mátyás Rákosi Steel Works applies the language of a thriller, suggesting that the waste collection campaign was so popular that it became a spontaneous, almost grassroots activity.

> The DISZ committee prepared a precise "schedule" detailing in which plant and at what time the youth were supposed to commence the metal search. They planned that on Thursday there would be general waste collection in the two most important shops: the assembly shop and the engine room. The successes of István Czutka's company, however, swept away this schedule like a deluge,[5] because, roused by their example, waste collection in the assembly shop as well as in the engine room began already Wednesday afternoon.
>
> It was one o'clock in the afternoon. The fitters, the locksmiths, and the welders were working at a calm pace in the middle assembly shop. In the workshop they were sweeping up, and in the semi-finished product store room the parts were lined up seemingly in the most complete order. One would think that in such a clean plant there would be no metal waste, that one could not find even a bad screw. After quarter past one a huge restlessness/commotion started in the whole workshop. The second year locksmith students, under the leadership of Imre Juhász, one of their own, began the metal collection on and under the work benches and behind the seat boxes. Older, more experienced workers directed every one of their steps, so as not to put some valuable material still to be worked among the iron scraps. They were making frequent trips with the wheelbarrows, and in one of the free corners of the shop the many metal sheets and other unusable iron materials were already piling high. The news that Imre Juhász, the leader of the small group, found a heap of copper at the leg of a box seat spread like wildfire. The shovel and a smaller box turned up within a minute, and Imre Juhász carefully shoveled even the last speck of copper dust into the small box. The result of the collection of an hour and a half was seven boxes of metal sheets, screws, and scrap. (*Szabad Nép,* May 19, 1951, 5)

To make the movement more attractive, the party decided to nominate a hero who could become, just like Stakhanov, the eponym of a new movement. That's when they picked Géza Gazda, a metallurgical worker employed by the Mátyás Rákosi Iron and Metal Works in Csepel, to lead the waste-reuse movement across the whole economy in 1951. Previously he had spawned a number of initiatives to collect and reuse metal wastes, and his approach was quickly embraced by the Budapest Party Committee and applied in all other branches.

The mobilization for the Gazda movement was even more total than for the previous material-saving campaigns, leading to the establishment of what practically amounted to a cult of waste. The cult of waste was embodied in the rituals of waste-collection weeks and offerings of collected wastes to the party or to Rákosi for his birthday; in the sacred objects of waste reuse, such as the Gazda movement idea boxes; and in the whole cast of waste-reducing, waste-reusing characters—from the role of the class-conscious waste stewards mobilizing their less conscious colleagues for a waste-related task to the waste-profiled brigades regularly challenging their counterparts to more and more waste reduction or waste reuse ideas. The early socialist discourse of waste established a whole new material culture.[6]

THE CONSEQUENCES OF THE EARLY SOCIALIST DISCOURSE ON WASTE

The dominant discourse on waste thus constructed waste as a secondary raw material, as something useful, and therefore the state's task was to make sure that appropriate use values were given to waste. Gazda's speech nicely illustrates this use value thinking: "What can you use a thousand tons of [scrap] reinforcing iron for? For the ferro-concrete ceilings of 750 two-room apartments, or for covering the iron-needs of three buildings as big as the Sport Stadium of Csepel, or . . . 50,000 bicycles or 64,000 sewing machines, or 160 large, radial drills, or boiler tubes for 300 engines" (Gazda 1951).

But if the usefulness of waste was clear, the usefulness of waste reuse was not. This discourse on waste, in fact, turned out to be counterproductive. First, because the reuse of waste usually required additional technical and raw material resources and labor input, mostly missing, the collected wastes were often left to rust and rot and became unusable. That this was happening was hinted at by the 2.500-21/1954 decree of the President of the Central Planning Office, which ordered that "[t]hose wastes that are usable within the Gazda movement without melting, breaking, etc. can be stored only in the amount that is sufficient for its processing for 90 days." This was also the case with MÉH, whose facilities, especially storing and sorting capacities, remained insufficient during the entire period of state socialism.

Second, reuse did not automatically save wastes from becoming trash. As economic considerations and public control were both lacking in decisions about waste reuse, the products made out of wastes often could not be used and, as a result, they degenerated into garbage. Reports about the achievements of the Gazda movement mostly summarized the results in terms of *potential* number of goods manufactured out of the waste or *potential* money savings.

> In the Fashion Shoe Factory, Miklós Szecsödi, . . . suggested making men's, women's or children's shoe vamps out of the wastes generated in the cutting of the colored welt, which before had been used to make the heel welt. This results in about 2-300 pairs of welts every month. (Leather Workers' Union 1951, 3)

> We can also expect serious results from the innovation of Béla Märcz, a worker of the Artificial Leather Manufacturing Company. The essence of the innovation is that while so far the rejects . . . could be made recyclable only through grinding, now even the smaller pieces are sown back into sheets, and this way there seems to be a saving of 19,000 Forints per year. (Leather Workers' Union 1951, 4)

It was only in a few cases that the transformation of waste into new goods was evaluated in the past tense, and in no case was it mentioned whether such goods had been actually bought or put to use. In 1959, the National Office of Materials and Prices was forced to introduce so-called "Gazda-movement prices," meaning discounts, which came as a final admission that these early recycled goods could not be sold at the regularly calculated consumer prices even despite shortages. However, as for the layaway supplies of Gazda-movement goods produced before 1959, for these neither the industrial ministries nor the Ministry of Domestic Trade could calculate and prescribe prices. What decades of economic reform could not carry through—a market-dependent price system—the Gazda movement achieved before anyone even thought about reforming central planning in Hungary:

> The industry [in this case the Ministry of Light Industries] cannot responsibly prepare a price calculation, nor has there been a controllable basis for the expenses in the price calculations in previous cases. The establishment of consumer prices is not feasible either; thus, the Ministry of Domestic Trade does not wish to deal with calculating prices for the products of the Gazda movement. . . . *Let this be the object of agreement between the producer and the trade companies*. (Hungarian National Archives, XIX-A-48. Országos Anyag- és Árhivatal. 524.1199/5. Ár-1959 [italics mine])

As collected wastes, whether recycled but not sold or not recycled at all, turned into trash, we can start seeing the truth in a Hungarian anthropologist's evaluation of the socialist effort to reuse waste as a "transformation of waste into garbage at significant expense" (Dömötör 1980).

Why then all the hue and cry about waste? Why the statistical questionnaires and surveys inquiring about waste generation and reuse? Why the Gazda idea-collecting boxes, why the Gazda exhibitions, why the brigades, why the stewards, why the waste-collecting weeks, why the pledges of new material savings made in honor of upcoming Party Congresses? Why all this enthusiasm and effort?

The party's key motivation for such total mobilization around waste was that, as is obvious from my previous discussion of the forms of waste, most wastes were produced haphazardly in amounts that were simply incalculable for the center. In this situation, the party-state, whose power depended on its ability to keep resources under central control and its legitimacy on its ability to redistribute, obviously had to make an effort to have firms produce some knowledge about their wastes. The aim of accumulating such knowledge was not necessarily to reduce their amount but to be able to control even these hidden resources and use their existence as an excuse to reduce input quotas for the constantly material-hungry firms, that is to keep plans even tighter.

The politics of waste proved to be a very effective tool for keeping in check workers, management, and private and cooperative producers and merchants. Documents indicate that knowledge of waste—its amount, its kinds, its potential uses and savings by reuse—was more important than making sure that wastes were economically reused, or at least that the goods thus produced were fulfilling real human needs, and even more important than ensuring that they were properly distributed. What an official complained about in an internal report was not the actual uses or squandering of wastes but about the lack of sufficient knowledge:

> The production ministries [meaning the ministries of heavy and light industries] are not clear themselves how much iron scrap the companies under their management use within

the Gazda movement, how much they transfer directly to firms of the local industry [meaning firms in cooperative ownership], and how much they are able to supply the waste-collecting enterprises. (Sarlós 1952)

The Hungarian bookkeeping system, while able to record the amount of waste returned to the storerooms, was completely unable to determine how much waste was produced in any production process because it included in the material input accounts even that portion of the raw materials that during manufacturing turned into waste. Again, this indicates that the exact knowledge mattered much less than inculcating the notion in economic actors that their use of the state's resources, even if they were wastes, were constantly under supervision.

Who was subject to this disciplinary politics of waste? Practically everybody who came in contact with wastes, because everybody could potentially divert wastes for unauthorized uses. Workers created household items out of scrap, which Miklós Haraszti (1977) in his ethnography of a Hungarian factory called "homers" and described as their only escape from the alienation of Taylorized work. This, of course, along with other appropriation of waste materials by workers, was considered theft. Other misuses by workers included selling wastes for money. Already in 1954 there were complaints at the party meetings of MÉH that some workers were privately selling the wastes bought by the company. No wonder that the most used quote from Lenin in waste-reuse propaganda materials was the one emphasizing that resources should be shared and used not according to the particular interests of relatives and acquaintances but according to the common interest of the entire society:

Communism begins where the simple worker starts to think, in a self-sacrificing way even as coping with hard work, about the increase of labor productivity, about saving *each and every pound of grain, coal, iron,* and other products even though those will not benefit him, nor his "next of kin" but "those distant ones," that is, the whole society. (Lenin, quoted by Mándi [1951, 15] and Ember [1952, 26])

Managers not only participated in such "misuses" of waste but, as evidence from the Soviet Union indicates, they even manipulated the concept of waste; they labeled perfectly good quality goods as waste so as to sell or trade them for inputs they failed to secure through formal channels (Berliner 1957).[7] It was exactly this commonality of interests between workers and management that the designers of waste politics tried to erode. The constant mobilizing of workers for collecting and reusing wastes infuriated managers, because they viewed such campaigns as bypassing their authority and as legitimating workers' neglect of their normal work tasks. The answer of an operative designer, one of the evaluators of the Gazda proposals in his factory, to the question of *Szabad Nép,* "why the Gazda movement was not developing in the Ganz Wagon Works," illustrates this attitude: "Oh, come on, the innovation fever will just go to people's heads" (*Szabad Nép,* October 10, 1951, 5).

But managers and workers were not the only objects of this disciplinary politics: so were the former independent craftsmen who, if not entirely nationalized, now had to be content with continuing their activities in a much restricted form.

In the plan for building socialism, cooperative and especially private forms of ownership were considered transitional; they were supposed to wither away, and the economic function of these transitional forms was restricted to providing the local population with various services and quality consumer goods. Obviously, these producers and merchants were treated as step-children of central distribution, who, if they wanted to survive, had to be content with materials and tools handed down from the state enterprises. As for what they produced, they were supposed to complement large-scale manufacturing by state enterprises, while their raw materials were supposed to be drawn from the wastes of the state sector (Elefánti 1953; KSH 1952, 1953).

Despite the declared aims for and the restrictions on the private and cooperative industrial sector, by the beginning of the 1950s it not only was thriving but started supplying state enterprises, paying preferential wages, and, by "neglecting the use of waste materials," more resembled large-scale manufacturing by state-owned factories than small-scale complementary services (KSH 1952, 1953). From the planners' point of view, all this was a disadvantageous development. Such an evaluation and complaints by various central organizations that the private producers and merchants transformed and/or resold at a profit layaway supplies and wastes bought cheaply from the state sector led to periodic central initiatives to revise purchase prices of waste materials for the private and cooperative sector, especially between 1951 and 1959. This attack, however, only opened up a can of worms. The two contrasting interests of the state—keeping a lid on the development of the cooperative and private sector on the one hand and the need for the utilization of the wastes of the state factories on the other—made regulations around waste use a contradictory business, ultimately leading to overregulation and occasional paralysis.[8] In general, the private and cooperative sector's economic activity was tolerated as long as it tapped the overflowing supplies of particular waste materials, but was looked at with suspicion when it competed with the state enterprises for wastes in high demand.

It is exactly because the reuse of waste became subordinated to such political considerations that the wastes already collected and even built into new goods failed to satisfy human needs and turned into garbage. However, there were two other not less important consequences of this politics: First, the Gazda movement, the campaigns, and especially the waste quotas all provided an added incentive to produce even more waste. To meet waste quotas, managers and workers had an explicit interest in producing with high waste ratios, and because of the ideology of the waste-reuse campaigns, they thought that such wastes would not be squandered but put to good use. Planners, as well as Gazda, warned against such an attitude, but the political and economic incentives for wasteful production, discussed above, and the propaganda, which in comparison to those few qualifiers that urged waste reduction as well, proved overwhelming and made it very clear for any individual what the priorities were.[9] This problem was admitted by central organs:

In 1956 those ministries, and the companies subordinated to them, that generate nonferrous metals stopped getting plan quotas. To wit, experience showed that it was impossible to prepare technically grounded and correct delivery plan quotas for the deliverers [of waste]. It happened that certain enterprises received quotas that were unrealistically high, which motivated them in the wrong direction, while other firms received loose delivery quotas, which they could fulfill much too easily. (Tájékoztató [a KGST-nek] a Magyar Népköztársaságban az ócska színesfém és hulladékgyüjtés megszervezésének megjavításával és a másodlagos színesfémkohászattal kapcsolatos kérdésekröl.) (Report [to the COMECON] on the issues of organizing and improving the collection of scrap nonferrous metals and of the metallurgy of scrap nonferrous metals in the Hungarian People's Republic. Országos Tervhivatal. [Central Planning Office] September 19, 1960. XIX-A-16-u. Box 32/a. Topic 5. Document Number 2-00188/1960)

Second, the most significant consequence of this discourse was that it denied legitimacy to the idea that waste may not be 100-percent recyclable or that waste could even be something harmful. The environmental consequences were disastrous.

As existing institutions kept pushing for waste reuse and avoiding the regulation of hazardous waste disposal, in 1967 the Executive Committee of the Council of Budapest was forced to admit that "the question of [the disposal of] industrial wastes is unsettled in administrative respects." As documents in local government and enterprise archives indicate, the problem was primarily seen as an obstacle to industrial development and as a problem of unplanned land use. Local government officials complained about illegal dumps appearing both next to dumps managed by the council and on the territories of recultivated full dumps, while a chemical firm warned the Ministry of Heavy Industry that

[t]he solution of the waste disposal problem is a rather urgent task, because the current uncertain and not always professional practice could cause a lot of damage to water management and impedes production and its development. The burying of waste by the chemical industry administration in the proposed manner [in abandoned mines] could guarantee the [planned] rate of development for a long period. (President of the Budapest Chemical Works, 1968)

Note that this company could not argue on environmental or even public health grounds, but instead had to appeal to the plan and especially to growth. In response to the firm's plea for legal dump sites, central authorities urged it to keep looking for waste-reuse possibilities.

But even if legal, centrally controlled solutions were lacking, enterprises did not fail to seek individual and somewhat independent solutions to their problems. The informal economic network that was thriving in the slowly thawing political atmosphere from the 1960s on offered itself as the obvious and probably the only terrain on which waste storage problems could be solved, at least temporarily. While the rationalization of cultivation freed lands, increased autonomy for state farms and cooperatives made it possible to find semilegitimate uses for these new derelict lands, that is selling or trading them as dump sites through the decentralized and increasingly informal decision-making channels.

While in the United States it is lands with the lowest market values and with the most impoverished, powerless, and mostly African American, Native American, or Latino population that end up hosting toxic waste dumps or incinerators, in socialist Hungary it was those lands that the state's political interests marked as somehow valueless for the building of socialism. Baranya, a county in the south of Hungary, for example, became a figurative wasteland even before the first hazardous wastes of distant towns and firms landed on its doorsteps. The industrialization of this county was neglected partly because of its "politically unreliable" ethnically mixed population (Croats and Germans) and partly because of its proximity to Tito's Yugoslavia. In addition, its hilly plots did not suit large-scale mechanical cultivation, which led to the cessation of production over large expanses and to the fusion of cooperatives, soon followed by the fusion of villages as well. The economically and administratively paralyzed villages were a perfect target for industries wanting to get rid of their wastes. In 1986, there were 48 waste dump sites registered in Baranya County, of which 25 were storing hazardous industrial wastes (MTESZ 1986). In comparison, Tolna and Somogy, the two neighboring counties which did not generate much less hazardous waste than Baranya, housed six and one hazardous waste dumps respectively.[10]

THE MONETIZED CONCEPT OF WASTE UNDER LATE STATE SOCIALISM (MID-1970s TO 1989)

The early period of socialism did not see waste dumping, but rather mass mobilization for collecting and reusing waste, coupled with a lack of public knowledge and control over actual waste reuse. While the institutions that controlled waste reuse remained intact, the central attention to waste production abated. There was no more legislation on wastes until 1981, and Gazda's name was soon forgotten. The easily accessible waste reserves, such as those found among the ruins of war and those lying around the yards of factories, as well as the enthusiasm of workers and youth troops exploring and collecting wastes in their free time, were exhausted (Dömötör 1980). It was also because technologies generating less waste were increasingly applied (Dömötör 1980). From the end of the 1960s, but more pronouncedly from the 1970s, the New Economic Mechanism had two contradictory effects. On the one hand, it made the enterprises' budget constraints harder, which supposedly had a beneficial impact on material efficiency and thus on the generated amount of waste. On the other hand, recycled products could not always fulfill the stricter quality requirements nor the new efficiency criteria that also came with the reforms (Dömötör 1980), which even if they were not real obstacles given the stop-go pace of the reforms, at least could be used as excuses for not reusing wastes.

The implementation of the reforms thus signaled the end of the first discourse on waste, which slowly gave way to a monetized concept of waste. Waste was conceptualized as inefficiency, and although it was still not seen as useless, its reuse was no longer expressed in use value or in fulfilling plans and building socialism. Rather, waste was seen as yet another cost of production and waste reduction and reuse as decreasing these costs. In addition, waste reuse was now subject to the same cost/benefit analysis as the manufacturing of other products, which called for a more selective collection of wastes.

An example is the 1981 Waste and Secondary Raw Material Management Program, which was implemented along with other rationalization programs in the early 1980s to combat the explosion of fuel and raw material prices on the world market. This program dealt with waste production merely because it helped reduce material expenditures in production. A Central Statistical Bureau report on the achievements of this program summarized its goals this way:

> The more rational management of material resources received a priority among the economic policy goals of the sixth Five-Year Plan. The program announced chronologically as the second among the austerity programs wished to achieve an increased substitution of primary raw materials with wastes, and the exploration and mobilization of economic reserves hidden in the recycling of secondary raw materials. It treated waste utilization as an activity reducing expenditures by which the per unit material costs can be reduced in relation to all costs of production. (KSH 1988, 10)

The shift in emphasis from the discourse prevalent at the time Gazda's movement started to take off is quite marked, as indicated by the following quote from a 1951 propaganda piece:

> [O]ur party has now been keeping the cause of material savings on the agenda for a longer time, but especially since the Congress. The leaders of our party have already emphasized many times that this is not simply talked about as one tool of decreasing production costs, but primarily as the basic condition for fulfilling the plans in many industries. (Ember 1952, 27)

Waste reduction and waste reuse now became professionalized: it was no longer the task of workers as in early state socialism, but rather the task of engineers and economists, both in central planning and in the higher management of enterprises. Reusing wastes now became the object of cost/benefit analyses, and funds were established to motivate firms to apply "waste-conscious" technologies. The state financed at least two-thirds of the costs of the reusing facilities, firms were exempted from fees on such investments, and 25 percent of the interest on loans borrowed for such purposes were remitted.

The way of fulfilling the task changed as well. In the first period of state socialism, waste reduction and waste reuse were to be achieved by making the most out of the existing production process, adding a few containers on nailing machines to collect falling nails or redesigning the tailoring of leather sheets. In the 1980s, however, the same goals were to be achieved by the total technological overhaul of production processes, prioritizing two types of investments: facilities to collect and prepare wastes for reuse, and capacities for reuse. The 1981 program made waste reduction and especially reuse a matter of technological development, and it equated these tasks with raising efficiency. However, because the system of distribution was left intact, the allocative form of waste prevailed. Between 1980 and 1985 the overall proportion of wastes in the composition of industrial products did not increase despite the 1981 Waste and Secondary Raw Material Management Program. In fact, in some industries it even decreased (Remetei 1986, 469).

While the monetized concept of waste had been gaining ground for a while, the 1981 Waste and Secondary Raw Material Management Program itself was, interestingly, a panicked reaction of the industrial "lobby" to the Decree on Hazardous Wastes that was put into effect in January that year (Takáts 1996). This decree defined hazardous wastes and regulated their elimination for the first time in Hungarian history, and it marked the official sanctioning of waste problems as environmental concerns. Even before 1981—as early as 1971—the waste problem was connected with environmental issues, but only to the extent of arguing that reusing wastes saved natural resources that were now admitted to be finite. It was in 1976 when the Hungarian media first started talking about wastes as potential polluting sources themselves and when the need to regulate hazardous waste disposal and treatment was first officially acknowledged. The most progressive aspect of the 1981 law was the paragraph that obliged large firms to record the material flow diagrams of their production processes, which made it easier for ministry officials to trace and estimate the quantity of the generated toxic by-products.

The waste-reuse program was limited inasmuch as it was mostly restricted to reuse, but the decree on hazardous wastes, with its exclusive focus on disposal, was not much more effective either. In practice, both ignored source reduction. In addition, both lacked guarantees of democratic control. Therefore the benefits of the simultaneous regulation of waste production and disposal, in theory an ideal situation, could not be reaped.

One tangible problem was the result of the state's insistence on the old notion of waste as useful material and on reuse as a solution to waste problems. While dumping now became professionalized, dumpsites were still created for temporary use only. That meant that even though these dumps were permitted and designed by several professional authorities, such as the health, geological, and water authorities, they were not meant to last long and to hold large amounts of waste. These sites also were considered temporary because the state hoped that in more prosperous future years the firms would be able to finance the reuse of wastes stored on them (Takáts 1996; Sztrevjopulosz 1996). As we now know, those years never arrived, and the temporary dumps turned into long-term wastelands, causing an environmental nightmare.

In sum, the second period of state socialism in Hungary was characterized by uncontrolled, nonprofit, professional, and regulated dumping. This is highlighted by the Dorog incinerator (operating since 1989) and the preparatory work for the first specifically toxic waste dump site in Aszód (operating since 1990). Participation and control were maintained for technocratic management only. There was moderate and unsuccessful public control over the establishment of the Dorog incinerator, but this was accidental as at this time institutional and legal guarantees for public participation were absent.

THE DOMINANCE OF DISPOSAL ISSUES IN THE TRANSITION (1989 TO PRESENT)

The transition is a period of publicly controlled, legally regulated, profit-oriented, professional dumping and of an exclusion of waste production from public discourse. The waste institutions of socialism are disbanded or become privatized and radically limited in their scope. The latter was the fate of MÉH, for example. The obligation of larger manufacturers to document the material flow of their production processes has been removed from the new environmental law. This signals the end of the state's control over the production and collection of wastes.[11]

Any survey of article titles coming out after 1989 could prove that the dominant perception of wastes now is that they are rather useless and harmful. Waste becomes entirely environmentalized, and to the extent that its production is being dealt with at all, it is reduced to a problem of efficiency. Mainstream policy studies still keep using per capita GDP indexes as a measure of environmental impact and thus treat the increase in efficiency and a change in the economic structure as an environmental panacea. The implicit suggestion is that squandering resources and polluting nature is appropriate, so long as it is done efficiently.

In general, waste distribution becomes the key focus of political and economic activities, ruling out concerns over production. To this extent, the transition strongly resembles core capitalism (Szasz 1994).

The already-existing Aszód waste dump and Dorog incinerator were privatized by French capital, and the infamous Garé dump is now about to be transformed into a toxic waste incinerator and dump of unburned deposits owned by a French–Hungarian joint venture. All over Eastern Europe there has been a tremendous "rush to burn," as Greenpeace calls the increasing trend of waste treatment by incineration. Since 1989, there have been an estimated 17.6 million tons minimum of annual incinerator capacity proposed just in Russia, the Baltics, Hungary, Poland, the Czech Republic, and Slovakia, with about 92 percent of this capacity offered for export by Western countries.[12] Put another way, there have been about 173 proposed facilities in the region. Austria is leading the way with participation in 28 projects, closely followed by Germany in 25, Denmark in 16, and the United States in 9.[13]

In the transition, there is no discourse on the production of waste, only on its distribution. Public participation is now legally regulated but limited to the distribution issue. The system of public hearings has been introduced in Hungary, which in several cases has proved to be quite a successful forum for resisting incinerators and waste dumps. But, as explained in the conclusion, this new politics of waste has its limits.

CONCLUSION

In conclusion, let me draw a very preliminary balance of the transition. So far, there is no evidence that material intensities dropped, and energy intensities never stopped growing, suggesting that Hungarian firms continue to produce relatively wastefully. The amount of hazardous wastes increased in the first year of the transition, then slightly decreased, but never returned to the 1989 level. Even this decrease, according to Hungarian experts, is due not to increasing efficiency but to the economic crisis and to a lack of responses to central surveys (Ministry of Environmental Protection and Regional Development 1994, 12). The risks from waste distribution don't seem to diminish either. The number of hazardous waste dumps increased by 12 percent between 1989 and 1994, and Western waste export schemes became more common as well. While the incinerators imported from the West might be thought to solve this problem, it is unlikely that Hungarian companies can pay for such services, which is why Greens argue that these facilities will end up importing waste from the West. In addition, the availability of such incinerators provides even less incentive to reduce waste at its source.

The political terrain seems a little bit more optimistic, but, as indicated above, the exclusive focus on distribution is a limitation. Szasz (1994) argued that in the United States the environmental-justice movement with its effective resistance to toxic waste facilities forced states to increase environmental standards; this increase, by making waste treatment more expensive, is now forcing firms to take measures for source reduction and prevention. The success of this model, however, depends on certain conditions. It needs firms that are sensitive to costs—that is, firms who, seeing the costs of dumping and incineration rising, will decide to seek the long-term financial benefits from source reduction. It also requires a strong state that can raise environmental standards so high that dumping and incineration costs would rise considerably. And it needs democracy to control all these activities.

In Hungary, most firms are now cost-sensitive, but many of them can still secure preferences through informal networks that softens budget constraints. Most importantly, however, the uncertainty of the economic future of any individual firm leads less to long-term than to short-term profitability considerations. Further, the state in Hungary is rather weak. The postsocialist liberal ideology that denies the state any agency prevents it from effectively arguing for and especially from enforcing a significant rise in environmental standards. Finally, local resistance is also weakened when it faces mostly foreign waste producers, since the feedback loops (going through residents, the local government or state, and the waste-generating enterprises) will of necessity be much more indirect than when facing domestic producers.

Ultimately, in this volatile situation some direct public control over waste production would be in order. However, as I argued, it is exactly such control that has been discredited by the socialist state, which even as it sought to salvage wastes, ended up reproducing them in ever greater volumes and preventing environmentally sound ways to deal with them.

NOTES

Research for this paper has been supported by the International Research and Exchanges Board, the Joint Committee on Eastern Europe of the American Council of Learned Societies and the Social Science Research Council, the Wenner-Gren Foundation for Anthropological Research, and the Project on European Environmental History, Sociology and Policy at the University of California, Santa Cruz. I am grateful to Michael Burawoy for helping me develop arguments in this paper.

1. See the debate between Sweezy (1971) and Binns and Haynes (1980) on the one hand and Ticktin (1976, 1992) on the other.

2. This was the assumption held by Eastern European reform economists (Szlávik 1991) and is now the hope of World Bank representatives (Manser 1993, 67).

3. This was the Party's daily; its title means "free people."

4. This finding begs the question whether this form of socialist wastefulness was due to the East European countries' position in the capitalist world system. Unfortunately, there are no data on other semiperipheral countries' waste generation to extend the comparison.

5. This composition echoes a very common phraseology of the October Revolution: "the revolutionary wave swept away the old social order," thus creating in the reader an unconscious continuity between the socialist revolution and the metal-collecting week.

6. The novelty of course lay not in assigning use value to waste, an attitude familiar to Hungarians due to long periods of deprivation, but in the state assuming this role and in politicizing waste reuse.

7. This is a common practice even after the collapse of central planning, although now it persists for different reasons (Burawoy and Krotov 1992).

8. In fact, there were two kinds of paralysis: the production of goods made out of these materials had to be stopped when access to these materials became limited, and the production process in which these waste materials were generated was hampered by their accumulation in factory yards.

9. Already the first regulation of waste recognized such a connection: "The collection ordered by the present decree must not interfere with the measures aiming at rationalizing material use, the innovations for waste reductions, . . . and it especially must not result in the increase of waste generated in production."

10. I only found data on these three counties, as at that time there were no central surveys of waste dumps. To put these data in context, there are 19 counties in Hungary.

11. This certainly goes for consumer waste, too. Most products are now sold in plastic, nonrecyclable containers, and supermarkets stopped collecting and recycling glass bottles.

12. My calculation from Greenpeace data (Gluszynski and Kruszewska 1996).

13. The country of origin of these incinerator schemes is not known in a large number of cases. In a few cases this is due to the fact that the bidding had not been completed at the time the data were collected. About 70 of these proposed facilities were stopped or withdrawn.

REFERENCES

Baran, Paul A., and Paul M. Sweezy. 1966. *Monopoly Capital.* New York: Monthly Review Press.

Berliner, Joseph S. 1957. *Factory and Manager in the USSR.* Cambridge, MA: Harvard University Press.

Binns, Peter, and Michael Haynes. 1980. New Theories of Eastern European class societies. *International Socialism* 2(7): 18–50.

Burawoy, Michael, and Pavel Krotov. 1992. The Soviet Transition from Socialism to Capitalism: Worker Control and Economic Bargaining in the Wood Industry. *American Sociological Review* 57: 16–38.

Castoriadis, Cornelius. 1978–1979. The Social Regime in Russia. *Telos* 38: 32–47.

Dömötör, Ákos. 1980. "A MÉH nyersanyaghasznositó tröszt története (1950–1980)" ("The History of the MÉH Trust [1950–1980]"). Budapest. Unpublished manuscript.

Elefánti, Ervin. 1953. Bevezetés (Introduction). In *Az anyagtervezés néhány problémája* (Some Problems of Material Planning). Kozgazdasagi Dokumentacios Kozpont. Budapest: Akademiai Kiado.

Ember, György. 1952. *Gazda-mozgalom a vasiparban* ("The Gazda Movement in the Iron Industry"). Budapest: Népszava, Szakszervezetek Országos Tanácsa Lap- és Könyvkiadó Vállalata.

Executive Committee of the Council of Budapest. 1967. Minutes from its session on November 22, Budapest.

Filtzer, Donald. 1992. *Soviet Workers and De-Stalinization: The Consolidation of the Modern System of Soviet Production Relations 1953–1964.* Cambridge, UK: Cambridge University Press.

———. 1986. *Soviet Workers and Stalinist Industrialization: The Formation of Modern Soviet Production Relations.* London: Pluto Press.

Gazda, Géza. 1951. "Használjunk fel minden gramm hulladékot!" ("Let's Use Every Gram of Waste!") *Szabad Nép* ("Free People"). August 14: 1.

Gluszynski, Pawel, and Iza Kruszewska. 1996. *Western Pyromania Moves East: A Case Study in Hazardous Technology Transfer.* Greenpeace. [http://www.rec.hu/poland/wpa/pyro-toc.htm].

Goldman, Marshall I. 1972. *The Spoils of Progress: Environmental Pollution in the Soviet Union.* Cambridge, MA: MIT Press.

Gomulka, Stanislaw, and Jacek Rostowski. 1988. An International Comparison of Material Intensity. *Journal of Comparative Economics* 12: 475–501.

Haraszti, Miklós. 1977. *Workers in a Workers' State,* trans. Michael Wright. London: Penguin.

Hughes, Gordon. 1990. *Are the Costs of Cleaning Up Eastern Europe Exaggerated? Economic Reform and the Environment.* London: Centre for Economic Policy Research.

Jócsik, Lajos. 1977. *Egy ország a csillagon* ("A Country on the Star"). Budapest: Szépirodalmi Könyvkiadó.

Juhász, Ádám. 1981. "Az anyagtakarékosság lehetöségei és feladatai az iparban—Tézisek," ("The Potentials and Tasks of Material Savings"). In: *Az információ 1981. évi helyzete—Az anyaggazdálkodás helyzete és fejlesztésének irányai* ("The Situation of Information in 1981: The Situation and Directions of Development of Material Savings"). The abbreviated version of talks given at the Ninth Itinerary Congress, Szolnok, Hungary.

Kornai, János. 1980. *Economics of Shortage.* Amsterdam: North-Holland.

KSH (Központi Statisztikai Hivatal—Central Bureau of Statistics). 1952. Report on the Activities of Cooperatives. Budapest: Központi Statisztikai Hivatal.

———. 1953. Report on the Activities of Cooperatives. Budapest: Központi Statisztikai Hivatal.

———. 1988. *Központi fejlesztési programok. A melléktermék- és hulladékhasznosítási program 1987. évi eredményei* ("Central Development Programs. The By-Product and Waste Reuse Program"). Budapest: Központi Statisztikai Hivatal.

Leather Workers' Union. 1951. *A Gazda-mozgalom elmélyítésével segítsük elö ötéves tervünk sikerét* ("Let's Foment the Success of our Five-year Plan by Deepening the Gazda Movement"). Budapest: Egyetemi Nyomda.

Mándi, Péter. 1951. A pártszervezetek feladatai az anyagtakarékosság terén ("The Tasks of the Party Organs on the Field of Saving Materials"). *Pártépítés* (Building the Party) (4): 15.

Manser, Roger. 1993. *The Squandered Dividend: The Free Market and the Environment in Eastern Europe.* London: Earthscan Publications. (Also published as *Failed Transitions: The East European Economy and Environment Since the Fall of Communism.* New York: New Press, 1993.)

Ministry of Environmental Protection and Regional Development. 1994. Magyarország veszélyes hulladékainak helyzete. Budapest: KTM.

Moroney, John. R. 1990. Energy Consumption, Capital and Real Output: A Comparison of Market and Planned Economies. *Journal of Comparative Economics* 14(2): 199–220.

Murphy, Raymond. 1994. *Rationality and Nature: A Sociological Inquiry into a Changing Relationship*. Boulder, CO: Westview Press.

MTESZ (Müszaki és Természettudományi Egyesületek Szövetsége) ("Alliance of Scientific and Technical Organizations"). 1986. *Hulladéklerakó Kataszter* ("Waste Dump Registry"). Pécs: MTESZ Baranya megyei Szervezete ("The Baranya County Organization of MTESZ").

Nove, Alec. 1980. *The Soviet Economic System.*. London: George Allen & Unwin.

Packard, Vance. 1960. *The Waste Makers*. New York: David McKay.

Reiniger, Róbert. 1991. Veszélyes hulladékok ("Hazardous Wastes"). *Anyaggazdálkodás Raktárgazdálkodás* 19(6): 1–7; (7): 14–19.

Remetei, Ferencné. 1986. A hulladék és másodnyersanyagok programjának eredményei és tapasztalatai 1981–1985 ("The Results and Experience of the Waste and Secondary Raw Materials Program"). *Ipari és Építöipari Statisztikai Értesítö* ("Industrial and Construction Industrial Statistical Bulletin") (December): 465–81.

Sarlós, Mihály. 1952. A vashulladék begyüjtés kérdései ("Issues of Iron Waste Collection"). Report for Internal Use of the National Planning Bureau. XIX-A-16a OT 446. doboz. 25 cs.f.sz. no file number. Új Magyar Központi Levéltár.

Schreiber, Helmut. 1991. The Threat from Environmental Destruction in Eastern Europe. *Journal of International Affairs* 44(2): 359–91.

Simai, Mihály. 1990. Környezetbarát fejlödésünk ("Our Environment-Friendly Development"). *Valóság* 9: 1–10.

Simons, Marlise. 1993. West Offers Plan to Help Clean Up East Europe. *New York Times*, May 4.

Solomon, Laurence. 1990. The Best Earth Day Present: Freedom. *Wall Street Journal* (Eastern Edition), April 20, A14.

Steichen, Girard C. 1991. Years of Grime and Grunge Will Take Decades to Clean. *Christian Science Monitor*, April 25, 11.

Sweezy, P. 1971. *On the Transition to Socialism*. New York: Monthly Review Press.

Szasz, Andrew. 1994. *Ecopopulism*. Minneapolis: University of Minnesota Press.

Sztrevjopulosz, Krisztoforosz. 1996. Interview by Zsuzsa Gille. Budapest.

Szlávik, János. 1991. Piacosítható-e a környezetvédelem? ("Is Environmental Protection Marketizable?") *Valóság* 34(4): 20–27.

Takáts, Attila. 1996. Interview by Zsuzsa Gille. Budapest.

Ticktin, Hillel. 1992. *Origins of the Crisis in the USSR: Essays on the Political Economy of a Disintegrating System*. Armonk, NY: M. E. Sharpe.

———. 1976. The Contradictions of Soviet Society and Professor Bettelheim. *Critique* 6 (Spring): 17–44.

Toffler, Alvin. 1970. *Future Shock*. New York: Random House.

Part III

10

Success and Impasse: The Environmental Movement in the United States and Around the World

Robert K. Schaeffer

The environmental movement that emerged in the United States achieved remarkable political success during the 1970s. It persuaded the government to adopt dozens of laws, created agencies charged with implementing domestic environmental policies, and made population control a goal of U.S. domestic and foreign policy.[1] Movement efforts to publicize environmental issues raised public consciousness and increased public participation in sound environmental practices and organizations.[2] Growing levels of public participation in environmental issues increased the membership and income of old and new environmental organizations.[3] What's more, the movement's theories, first advanced in the 1960s, anticipated events in the 1970s and served as useful guides to the movement's organizing efforts.

Given its rapid and remarkable successes in the 1970s, many observers agreed with Robert Nisbet's 1982 assessment: "It is entirely possible that when the history of the twentieth century is finally written, the single most important social movement of the period will be judged to be environmentalism" (Dalton 1994, 243). But Nisbet's assessment may have been premature.[4] After Ronald Reagan took office in 1981, the movement was unable to pass new laws or defend earlier legislation, and its political progress stalled. The movement's political problems were compounded by the failure of some of its theoretical expectations in the 1980s, and by its inability to lobby, litigate, and organize effectively in new political circumstances. After 1990, as the movement's political weakness became apparent, public support for environmental organizations waned and membership plummeted, creating a crisis for many groups. The movement's political impasse and organizational crisis has generated an acrimonious public debate about the social character and political direction of the movement. In this context, it is important first to explain how the U.S. environmental movement successfully changed state policy, raised public consciousness, increased its domestic organizational strength, and spread its ideas to some other countries around the world in the 1970s. It will then be possible to explain why the political fortunes of the environmental movement in the United States and in other countries changed in the 1980s and 1990s.

MOVEMENT SUCCESS IN THE 1970s

The environmental movement's success in the 1970s was made possible by four important social, economic, political, and theoretical developments. First, the union of diverse social groups in the late 1960s and early 1970s created a social movement that could promote change.

The environmental movement has often been described as consisting of two parts: conservationist and environmentalist (Mitchell 1991; Constain and Lester 1995, 26). The conservation movement, which included turn-of-the-century groups like the Audubon Society and Sierra Club, was joined in the late 1960s and early 1970s by new environmental groups like the Environmental Defense Fund, Natural Resources Defense Council, Environmental Action, Friends of the Earth, and Greenpeace.[5]

These two currents had different origins and orientations. The conservation movement can be traced back to the British and American movements that sought to protect animals, children, slaves, and sailors in the 1830s and 1840s.[6] These movements assumed responsibility for protecting domesticated animals, children, slaves, mariners, and later birds (Audubon) and wild lands (Sierra Club) because these groups had no legal standing and could not effectively "protect" themselves.[7] So these movements assumed a "custodial" role, acting to protect animals, people or nature from abuse, much as guardians do for "minors" (Glenn 1983, 413). Conservation groups sometimes differed over the extent and kind of protection they might offer nature—John Muir favored preservation and Gifford Pinchot advocated conservation—but they shared a custodial approach. Still, it was not until the early 1970s that a series of court decisions gave conservationists the legal standing they needed to act effectively as custodians, to exercise what they called "stewardship" (Andrews 1980, 227, 231; Mitchell 1991, 10102; Turner 1990, 16–21). These rulings made it possible for conservationists to litigate on behalf of their wards, an important tactic for the movement in the 1970s.

By contrast, participants in the new environmental movement did not see themselves as guardians. They were instead concerned about how environmental problems affected them and insisted on protecting human "rights," such as the right to clean air. So while conservationists had an altruistic orientation, environmentalists adopted a "self-interested" perspective. These different approaches to the natural world or "environment" grew out of different social circumstances. In the 1960s, conservation groups were supported by elderly, upper-class Republicans, a testimony to the conservationist traditions established by Teddy Roosevelt and the Progressives during the early years of this century. The early environmental groups, by contrast, were supported by young, middle-class Democrats, whose interest in the environment grew out of a concern for civil rights and participation in the student, antiwar, and women's movements of the 1960s (Scheffer 1991, 6–7; Milbrath 1984, 75–76). The organization of Earth

Day by the elderly Republican Senator Gaylord Nelson and the young student activist Denis Hayes personified the two different constituencies that joined to create the "environmental movement" of the 1970s.[8]

Many writers have described the environmental movement as the union of conservationists and environmentalists. But few have explained why this union was forged at a time when age, gender, political, and social divides were sharply felt. In the late 1960s, one might have expected the new environmentalists to reject an alliance with the old conservationists and proceed alone, much like the New Left, which broke with the Old Left in the early 1960s (Teodori 1969, 2–54, 163–239; Gitlin 1987, 45–80). But conservationist and environmentalist groups did not go separate ways because social and political differences were bridged by shared experiences, common understandings, and mutual goals. The postwar baby boom gave both young and old a first-hand experience with rapid population growth and of the problems of scarcity associated with it.[9] Paul Ehrlich's argument that population growth was an environmental and social problem found support among both old and young, helping to bridge the "generation gap" that otherwise loomed large in this period. Mutual agreement on the need for population control, reproductive rights, family planning, birth control, and abortion (environmental groups like Zero Population Growth filed amicus briefs in *Roe v. Wade*) drew women into the movement and bridged the gender gap, which was then dividing other contemporary movements (Luker 1984, 142–43). And the spread of higher education, which enabled wealthy conservationists and middle-class students to discuss issues as intellectual equals, helped bridge class divides, which had loomed large when college education was restricted to the rich.

Conservationists and environmentalists also shared mutual political goals. They recognized that political success might be possible if they worked together, not separately (Andrews 1980, 239). Their common desire for change helped them put aside the ideological differences associated with their separate party identifications—Republican conservationist and Democratic environmentalist—and fashion a bipartisan approach to the state, which in any case was divided between a Republican executive and a Democratic Congress (McClosky 1992, 78; Pulido 1996, 24). This political collaboration was possible in part because conservation groups had not been compromised by the kind of close association with state power that had made collaboration between the Cold War Old Left and the antiwar New Left extremely difficult. As a result, the conservation movement could supply experience and money to the environmental movement and gain members, enthusiasm, and new issues in return.[10] These developments made it possible for diverse social groups and political forces to unite in a diverse movement that reached across age, gender, class, and political divides (Andrews 1980, 222–24; Buttel and Larson 1980, 326; Costain and Lester 1995, 26; Dunlap and Mertig 1992, 5; Schnaiberg and Gould 1994, 150; Scheffer 1991, 67).

Second, economic developments, particularly inflation, provided a material basis for environmental concerns and contributed to the movement's political success in the 1970s. During the late 1960s and early 1970s, inflation became an important economic problem. In 1972, President Nixon imposed wage-and-price controls to curb inflation, then increasing at about 4 percent annually (*New York Times,* August 16, 1971; Schaeffer 1997b, chap. 4). His efforts failed because the OPEC oil embargo and Soviet grain shortages increased prices dramatically and accelerated inflation into double-digit annual rates after 1973. Higher energy, food, and natural resource prices made these goods "scarce" for populations in the United States and the core. Increasingly, scarcity made credible the environmental movement's economic argument that energy, food, and natural resources were diminishing and that there were real "limits to growth." As one environmental writer in 1980 argued, "The inevitability of resource scarcity is a staple in modern environmental thought. Thanks to OPEC it is fair to say that these ideas are now much more widely appreciated by the public than they were in the early 1970s."[11] Under conditions of apparent "scarcity," people began using renewable energy sources and practicing conservation, which reduced energy demand for the first time in the postwar period. By making goods scarce, inflation corroborated the environmental movement's economic arguments and contributed to its political success.

Third, President Nixon and the Republicans in the House and Senate played a crucial role in the environmental movement's political success. Movement historians have discounted Nixon's role, arguing that he initiated, supported, and signed environmental legislation only because he was cynical, opportunistic, or simply indifferent to these issues.[12] They argue that Nixon supported environmental legislation to steal the increasingly popular issue from Democratic Senator Edmund Muskie, who had emerged by 1970 as Nixon's most likely challenger for the presidency (DeWitt 1994, 23–24; Dunlap 1992, 91–92; Rathlesberger 1972, viii). But this misrepresents Nixon's motives and minimizes his role, which was rather more substantial than movement historians admit.

Although Nixon's initiatives—creating the EPA by executive order, establishing the Council on Environmental Quality, banning DDT, adopting population control as a domestic and foreign policy, and signing major environmental legislation—are often depicted as attempts to "steal" the issue from Muskie and the Democrats, they were more an effort by Nixon to prevent Muskie from "stealing" these issues from Republicans, who had long expressed support for conservation and population control. Nixon wanted to prevent a repeat of 1960, when John Kennedy "stole" the civil rights issue from Republicans. Nixon acted in 1970 to ensure that Muskie would not likewise steal environmental issues from the Republican party in the upcoming elections (Branch 1988, 305–08, 321–23, 341–42, 348–75). Nixon saw himself as belonging to a Republican conservationist tradition stretching back to Teddy Roosevelt. As Nixon argued in 1970, a few months before Earth Day:

> At the turn of the century, our chief environmental concern was to conserve what we had—and out of this concern grew the often embattled but always determined "conservation" movement. Today, conservation is as important as ever—but it is no longer enough to conserve what we have; we must also restore what we have lost. We must go beyond conservation to embrace restoration.[13]

And Nixon's close association with the Rockefellers exposed him to their enthusiasm for population control, an important issue for John D. Rockefeller III, who had organized the Population Council in 1952 (Mass 1976, 37; Pope 1972, 170; Teitelbawm 1992/93, 66). Nixon initiated domestic family-planning programs, made population control a feature of U.S. foreign policy, and, at the first U.N. Conference on Population, "put forward a forceful policy agenda urging global action to reduce high fertility, including demographic targets" for every country (Pope 1972, 163–79; Teitelbawm 1992/93, 67). Nixon's support for population control, then a central tenant of the new environmental movement, contrasted sharply with that of many Democrats, who drew support from working-class Catholic constituencies.

Nixon's presidency was characterized by bold initiatives: breaking-up the international monetary system, opening diplomatic relations with China, exiting from Vietnam, initiating a war on drugs, and imposing wage-and-price controls (Schaeffer 1997, chap. 10). In this context, Nixon's environmental initiatives were serious attempts to grapple with important problems, and these initiatives often anticipated or exceeded the demands of movement activists and their congressional allies (DeWitt 1994, 24; Rathlesberger 1972, xii–ix). Although the movement is reluctant now to recognize Nixon as the most "environmental" president since Teddy Roosevelt, the fact remains that the movement's most important political gains came during his presidency. But while the movement's alliance with Nixon and its bipartisan work with Congress produced major *domestic* gains, the movement's close association with the U.S. state and its foreign policy—particularly the state's support for population control—antagonized states in the periphery and prevented, for a decade, the spread of environmental ideas to the periphery.

Environmental-movement ideas and organizations did not spread to the periphery during the 1970s for different reasons. First, many in the periphery were antagonized by the environmental movement's close association with the U.S. state, which they regarded as imperialist, and by U.S. population-control efforts. They were particularly troubled by the fact that the Nixon administration increased spending on the U.S. Agency for International Development's population control program while cutting its funding for overseas health care, tying food aid to a country's population-control efforts, and, after 1974, reducing food aid generally. In this case, U.S. policymakers used Ehrlich's concept of "triage" (withholding food aid from countries like India that were deemed incapable of becoming self-sufficient in food) as their rationale for changed food aid and population control policies (Ehrlich 1968, 159–61; Bandarage 1994, 294; Mass 1976, 48, 50–58, 138).

Second, global inflation had a different economic meaning in the periphery. For producers of energy, food and natural resources, inflation meant "plenty," not scarcity. Rising prices in the 1970s generally increased incomes in the periphery, and the transfer of wealth from the core to OPEC created surpluses that were used to finance economic development throughout the periphery (Schaeffer 1997b, chaps. 4 & 5). Changed terms of trade fueled an optimism in the periphery that population growth and economic growth might go hand-in-hand. These developments made environmental ideas about population control and the limits to growth unwelcome in the periphery. A few environmental groups did organize in Malaysia and Sri Lanka, largely as a result of Friends of the Earth's efforts. But because dictators ruled many states in the semiperiphery and periphery, dissident environmental groups found it difficult to organize or raise environmental issues effectively (Sardar 1981, 6). India was perhaps the exception during this period. Indira Gandhi's coercive population control policies, large-scale development projects, and emergency political rule antagonized many local groups, giving rise to widespread environmental activism that could draw effectively on grassroots and democratic political traditions in India (Tucker 1981, 10–11).

But while the U.S. environmental groups did not export their ideas to the periphery in the 1970s, their ideas and organizations did spread in Western Europe. The environmental movement in Europe grew in part because U.S.-based movements—particularly Friends of the Earth and Greenpeace—organized effectively there, transferring money, ideas, and tactics to sister or member organizations in Europe.[14] The 1972 U.N. Conference on the Human Environment in Stockholm played an important role by calling European attention to environmental issues, much as the 1992 U.N. Conference on the Environment in Rio would later call attention to environmental issues in the periphery (Caldwell 1992, 64; Lyons 1988, 21). European movements also grew out of the New Left. For the most part, European environmentalism was not a union of "conservationist" and "environmental" movements because conservationist organizations outside of the United Kingdom were rare (Prendiville 1994, 106). Instead, most environmental groups in Europe grew out of the political New Left and the relatively apolitical counterculture (Dalton 1994, 64; Lyons 1988, 17; Prendiville 1994, 106). Participants in the New Left began organizing "Green" political parties in Europe during the 1980s, while participants in the counterculture organized less political "lifestyle" environmental groups (Switzer 1994, 37). For example, Brice LaLond once described French environmentalism as an "epicurian ecology—the ecology of enjoying life, having animals, drinking and loving and talking, getting cheese and dessert and a little bit more, as we say."[15] The New Left and counterculture in Europe were joined in the 1970s by their common experience with inflation and scarcity and by their opposition to nuclear power and weapons. For example, all of the Friends of the Earth sister organizations worked against nuclear power in the 1970s and against nuclear weapons in the 1980s (Warner and Kreger 1984). As West German Green Party representatives explained in 1983, the Green Party coalition, which brought together remnants of the New Left and the counterculture, was "sewn together with a hot needle," the hot needle being nuclear power and nuclear weapons (Schaeffer 1983, 12).

Although environmental movements emerged throughout much of the core in the 1970s, they did not flourish in Japan. In 1990, for example, Greenpeace counted only 300 members in Japan, compared to more than 2,000,000 in the United States, 80,000 in the Netherlands, and 7,000 in France.[16] Environmental issues did not find organizational expression in Japan largely because the Japanese were greatly antagonized by the U.S. environmental movement's efforts to use the International Whaling Commission to curb Japanese whaling (Switzer 1994, 42). Anger about this issue made it extremely difficult to organize support for other environmental issues in Japan, much as anger about U.S. population-control policies made it difficult to raise environmental issues in the periphery.

Fourth, the environmental movement's political success was assisted by an astute set of theories or sets of expectations about the future. Because the expectations advanced in theories developed in the 1960s were apparently confirmed by events in the 1970s, they contributed to effective movement politics.

Generally speaking, the major theories adopted by the environmental movement in the 1960s, 1970s, and 1980s were developed by authors of popular books—Rachel Carson's *Silent Spring,* Paul Ehrlich's *The Population Bomb,* Amory Lovins' *Non-Nuclear Futures,* Jonathan Schell's *The Fate of the Earth,* and Bill McKibben's *The End of Nature*—many of which first appeared in *The New Yorker.*[17] Although they addressed different topical problems—pesticides, population, nuclear power, nuclear war, and global warming—they shared a common purpose: to anticipate and avert some awful environmental problem.[18]

Although many critics have commented on the movement's apocalyptic character, they have not explained why this should be a salient feature, not only for environmentalists but for religious fundamentalists in the same period (Roszak 1992; Schneider 1994). The apocalyptic literature of millennial Christians runs parallel to that of environmentalists.[19] And Michael Barkan has argued that environmental and religious literatures are both

> conducted in the idioms of social and political criticism . . . [evoking] world destruction and transformation through ecological disaster, nuclear holocaust and technological breakdown. The curious result is that two quite different bodies of apocalyptic literature flourish simultaneously . . . they converge on the belief that the accepted texture of reality is about to undergo a staggering transformation, in which long-established institutions and ways of life will be destroyed. (Boyer 1992, 336)

For whatever reason, doomsday theories have found receptive audiences in both secular and religious communities. But while doomsday theories have been used to mobilize both communities, the theories advanced by secular environmentalists were more readily tested by real-world developments.

During the 1960s, Carson, Ehrlich, and Lovins advanced theories that were adopted by the environmental movement or important parts of it. In *Silent Spring,* Carson expected three things to occur as a result of widespread pesticide use. First, she argued that pesticides would kill not only harmful insects but also "nontarget species" such as insect predators, birds, household pets, and humans. Second, she maintained that some insects would survive the chemical onslaught and then thrive, making pesticides increasingly ineffective as insect "resistance" grew stronger. And third, she argued that poisons would spread in the environment and pose a growing threat to species at the top of food chains. Her expectations, which were based on a scientific literature then in its infancy, were subsequently confirmed by scientists. During the next two decades, scientists found that nontarget species were harmed, pesticide resistance increased, and poisons spread, posing problems for animals and humans.[20]

Carson's arguments also provided a useful guide to grassroots antipesticide groups in the United States, which were grouped in the National Coalition Against the Misuse of Pesticides (NCAMP). The 1981 publication of David Weir and Mark

Shapiro's *Circle of Poison,* which found that DDT use continued in the periphery after it was banned in the United States and that DDT-contaminated produce was frequently exported to the United States, led in 1982 to the formation of the Pesticide Action Network (Schaeffer 1985, 10). In the 1970s and 1980s, grassroots groups, guided by Carson's theories and supported by a growing scientific literature, found that where pesticide use was controlled by local authorities—by city, county, or state officials, or by private utility companies, railroads, or farmers—they could litigate, lobby, and protest successfully against pesticide use.

In *The Population Bomb,* Ehrlich expected four things to occur as a result of rapid population growth. First, he argued that the population would outstrip the available food supply, leading to widespread hunger and famine in the mid-1970s. Second, he maintained that food shortages would lead to conflict and war. Third, he warned that a growing population would increase levels of pollution and waste. And fourth, he argued that the growing population would rapidly deplete nonrenewable resources (Ehrlich 1968). To forestall these developments, Ehrlich advocated aggressive population-control measures and recommended that the concept of "triage" be used as a way to determine which countries should receive food aid and which should be denied aid (Ehrlich 1968, 159).

Events in the 1970s seemed initially to confirm Ehrlich's theoretical expectations. The 1972 famine and war in Bangladesh, Soviet grain shortages, the energy crisis, and inflation all seemed to indicate that population growth was related to hunger, war, pollution, and shortage.

In *Non-Nuclear Futures* and *Brittle Power,* Amory Lovins advanced a simple expectation about nuclear power: accidents happen. Lovins argued that complex technologies were "brittle" or accident-prone. Lovins warned that the risks of failure were magnified when a brittle technology was combined with nuclear materials. He noted that the industry's attempts to reduce risk by adding safety features would be counterproductive because they would make the technology more complex and therefore more prone to failure, and they would greatly increase the cost of producing energy, making it uneconomical.

Lovins's expectations were made credible by a series of small-scale accidents and snafus that came to light as a result of diligent research by antinuclear activists and information provided by industry dissidents and whistle-blowers (see the regular column "Nuclear Blowdown" by Jim Harding, *Not Man Apart*). Lovins's theoretical expectations proved a useful guide to grassroots antinuclear groups across the country. These groups were successful for a variety of reasons. First, authority for nuclear power had been delegated by the U.S. government to public utilities and private businesses, which made them vulnerable to grassroots litigation, lobbying, and protest. Second, government laws about nuclear power were less stringent than those applying to nuclear weapons, which made grassroots protest easier. Third, the energy crisis and inflation slowed the demand for energy and increased the cost of nuclear power. Rising costs and the risks associated with nuclear power made it seem an increasingly poor investment for utilities and stockholders. This confirmed Lovins's argument that nuclear power was "uneconomical." Finally, the partial meltdown at Three Mile Island in March 1979 confirmed antinuclear arguments that major accidents could happen (Komanoff Energy Associates 1992; Schwab 1994, 221).

THE IMPASSE OF THE ENVIRONMENTAL MOVEMENT

The environmental movement's political success did not long endure. Beginning in the 1980s, the movement reached a political impasse. It failed to pass new legislation or improve old legislation that came up for renewal. Although membership and financial contributions increased during the 1980s, environmental organizations experienced sharp declines in the 1990s as the movement's political ineffectiveness became apparent (Schneider 1995). In 1994, *U.S.A. Today* reported that the ten largest environmental groups "were facing their worst financial and philosophical unrest ever, with membership down 6.5 percent and income flat since 1990."[21]

The movement's political impasse has generated an acrimonious debate about its social character and political practice. Critics blame the impasse on the movement's "elitist" and "upper middle class" social character and on the "professionalization" of the major organizations (Devall 1992, 55; Dowie 1996, xi–xii, 146–47, 205–06; Grossman 1994, 291; O'Rourke 1992, 22; Pulido 1996, 23; Rushefsky 1995, 280; Schneider 1995; Seager 1993). They argue that these "eight-to-five" or "reform" environmental organizations have sought political compromise and accommodation with the state, which has weakened the movement's ability to promote real change (Bloch and Lyons 1993, 7; Devall 1980, 303, 319–20; Devall and Sessions 1985, 3; Lewis 1992, 7–8). Other critics argue that many grassroots organizations promote an "eco-extremism" that undermines the credibility of "progressive" environmental groups and makes responsible change difficult to achieve (Lewis 1992, 1–2).

The movement's political practices are also the subject of debate. Critics debate the effectiveness of lobbying as opposed to grassroots activism and argue that the movement devotes attention to the wrong or insignificant issues (recycling is widely criticized as insignificant), which deflects attention from the really serious issues (Devall 1980, 299; Devall 1992, 51–52; Easterbrook 1990; Kaufman 1994, 5, 7; Lilienfeld and Rathje 1993; Schnaiberg and Gould 1994, v; Stevens 1993; Tierney 1996). While most critics agree that attention should be paid to the serious problems and that the movement risks "courting irrelevance" if it does not, they disagree about what the serious issues are and how they should be collectively addressed (Devall and Sessions 1985, ix; Dowie 1996, 1, 144; Schnaiberg and Gould 1994, 161; Schwab 1994, 417–18). These debates grow out of a frustration with the movement's political failures and generally reflect the political differences of the movement's two social constituencies: conservationist and environmentalist. Rather than arbitrate these debates and assess their merits, I want to analyze how political, economic, theoretical, and social developments in the 1980s and 1990s changed the movement's political fortunes in the United States and around the world.

First, important political developments contributed to movement impasse. After Reagan took office in 1981, his administration cut budgets for environmental programs and policies, eliminated regulations designed to protect the environment, refused to enforce environmental legislation, transferred 20 million acres of public land to private owners, and appointed hundreds of conservative judges to the federal bench (DeWitt 1994, 53; Dowie 1996, 78–80, 92; Kreger and Warner 1984, 116; Resenbaum 1995, 224; Scheffer 1991, 18; Schneider 1992). In effect, the Reagan administration disassociated state policy from environmental-movement influence. As Interior Secretary James Watt told Sierra Club President Michael McCloskey: "We're going to get things fixed here, and you guys are never going to get it unfixed when you get [back] in" (Kreger and Warner 1984, 10).

Not only did Reagan evict the environmental movement from the state, he also drove congressional conservationists from the Republican Party. Support for environmental legislation within the Republican Party declined sharply during the 1980s, as the party denied its own political traditions and purged conservationists from its ranks (Kamieniecki 1995, 156). These two developments destroyed the bipartisan basis for environmental legislation on Capitol Hill, which had been so effective in the 1970s. And it drove the environmental movement into the Democratic Party.

Initially, Reagan's antienvironmental onslaught, which McClosky said "stunned the movement," helped environmental groups mobilize (McClosky 1992, 81). Leaders of the major environmental groups agreed to collaborate in the Group of 10, organized a petition drive to oust Secretary Watt, and created Political Action Committees to contest the Republicans in the 1984 elections (Dowie 1996, 68–69; McClosky 1992, 82). During the 1984 election, groups like the Sierra Club and Friends of the Earth for the first time endorsed a presidential candidate: the Democrats' Walter Mondale (Ingram, Colnic, and Mann 1995, 134–35).

But by abandoning its bipartisan political approach and joining the Democratic Party, the environmental movement tied its political fortunes to a party whose presidential candidates were crushed in general elections by Reagan and then Bush. As a result, the movement became associated in the public mind as just another "special-interest" group within the Democratic Party, while the party's political defeats crippled the movement's ability to lobby Congress effectively. The growth, meanwhile, of antienvironmental grassroots groups in the Wise Use movement, which served as Republican auxiliaries (alongside the religious right) outside the Beltway, created a constituency that could counter the environmental movement's pressure on local, state, and federal representatives and could litigate effectively against environmental laws in the increasingly conservative courts (Schneider 1992).

While the environmental movement experienced a general defeat in the 1980s, it nonetheless won some important battles. Antitoxics grassroots groups grew dramatically and enjoyed considerable political success in the 1980s despite, or rather because of, the Reagan administration's policies.

The antitoxics movement emerged in poor and working-class communities across the country after Love Canal in 1978 (Bullard 1994; Christrup and Schaeffer 1990; Dowie 1996). Generally speaking, grassroots groups emerged in response to the discovery of chemical hazards in their communities. The women who typically led them organized protests against landfills, dumps, incinerators, and chemical use in rural and urban settings (Bullard 1994, 100; Seager 1993, 26970, 275; Szasz 1994, 70). By 1984, 600 groups participated in the Citizen's Clearing House for Hazardous Waste (CCHW), an umbrella organization that services many grassroots groups, and 4,687 groups participated in antitoxics campaigns by 1988 (Szasz 1994, 72). These groups successfully blocked the construction of nearly 100 garbage incinerators; banned aerial pesticide spraying in many counties; forced companies to install pollution equipment, withdraw pesticides from the market, and clean up waste dumps; raised public awareness (most Americans list hazardous waste as a high environmental priority); and passed laws requiring states to ban or curb toxic chemical use (Christrup and Schaeffer 1990; Freudenberg and Steinsapir 1992, 33–34).

Antitoxics grassroots groups made political gains in the 1980s for a couple of reasons. First, they were determined opponents of toxic chemicals. When chemical hazards were discovered in poor communities, property values typically plummeted. Because poor and minority residents found it extremely difficult to flee the hazard, they fought tenaciously in defense of their homes and communities.[22] Second, their efforts were inadvertently assisted by Reagan administration policies. During the 1980s, the Reagan administration dismantled the EPA, thereby weakening its control over pesticide and toxic-use policy. It also delegated considerable federal authority to state and local governments as part of its devolutionary "New Federalism," which was designed to reduce federal authority and promote states' rights (Kriz 1989, 2991; Lester 1995, 41–42). The Reagan administration expected that weakened federal authority would make it easier for industry to operate without restraint. But grassroots groups found that dispersed local authority was more vulnerable to protest than central federal authority and that organized protest could be effective in contests with local government and private industry. What's more, the conservative, Reagan-appointed judiciary often sided with grassroots groups in court, agreeing that local authorities had a right to regulate pesticide and toxic chemical use. In 1984, for example, judges appointed by Governor Reagan to the California Supreme Court ruled that Mendocino County could regulate pesticide use, a decision that NCAMP spokesman Jay Feldman said was "extremely important" because it meant "that local communities have the power to protect themselves from the effect of pesticides" (Schaeffer 1984b, 5). After many favorable rulings in courts across the country, the Reagan administration recognized the risks of devolution and attempted then to reassert federal authority and centralize control over toxic chemical use in the EPA (Schneider 1992; Shabecoff 1989). Some of the major environmental groups initially supported the Reagan administration's recentralization efforts to grant federal preemption of local regulations because they wanted to restore the EPA to it previous status. But Lois Gibbs' CCHW and local grassroots groups were able in 1986 to block the administration's proposal and persuade the major environmental groups to withdraw their ill-considered support after a furious campaign (Dowie 1996, 136). By taking advantage of the political and legal opportunities that were made available to them, determined grassroots groups were able to achieve real success in the 1980s.

When the Democrats recaptured the presidency in 1992, environmentalists expected their political fortunes to improve (Buttel 1995, 187). But while

Clinton appointed some environmentalists to positions in the administration, the movement's support for the party produced few tangible results (Dowie 1996, 119, 179; Gilliam 1995). The movement's impasse continued because the Clinton administration focused on the economy, health care, and free trade agreements during its first two years, neglecting environmental issues.[23] The movement's position within the Democratic Party made it hard for it to criticize the administration, and the movement was disarmed by the expectation that the administration could be counted on to take care of the environment. Much the same thing had happened after the Democrat Jerry Brown replaced Reagan as governor in California. As the Sierra Club's Carl Pope observed, "The [environmental] constituency lost its momentum. [Because the movement got] the subliminal message that 'some one else is taking care of things,' constituents stopped writing angry letters. . . . Two years after Brown was elected . . . he turned to his former campaign manager and asked, 'Are there any environmentalists left in California?'"[24] Environmentalists also found that other Democrats opposed environmental policies and blocked political appointments, making it difficult to make gains even though Democrats controlled the executive and the Congress (Dowie 1996, 182, 184, 186). Of course, after Republicans took control of Congress in 1994, the brief prospect of political gains evaporated. As a result, the movement found that it was difficult to "unfix" what Secretary Watt and successive Republican administrations had previously "fixed" (Kreger and Warner 1984, 10).

Second, economic developments in the 1980s undermined support for the environmental movement in the core, while helping expand the movement in the periphery. In 1979, Paul Volcker and the Federal Reserve used high interest rates to curb inflation in the United States (Schaeffer 1997b, chap. 4). This policy reduced inflation in the United States, but it triggered a debt crisis for dictatorships in the periphery that had borrowed heavily in the 1970s (Schaeffer 1997b, chap. 5). The debt crisis of the 1980s contributed both to structural adjustment and to democratization (Schaeffer 1997a, chaps. 5 & 6). And because indebted states increased their exports of energy, food, and natural resources to earn income that could be used to repay debt, commodity supplies increased and prices fell.

In the core, lower inflation rates and falling commodity prices meant an end of "scarcity" and a return of "plenty." Under these conditions, U.S. industries and the public abandoned "conservation" and embraced consumption. As an editorial in *Not Man Apart* argued in 1983, "Lower oil prices may undo much of the progress that's been made toward developing alternative sources of power. No longer driven by necessity, the technological push toward [alternative] sources of energy will slow drastically" (Kaiser 1983, 2). That is exactly what happened: U.S. production of energy from solar, geothermal, wind, waste, wood, photovoltaic, and solar thermal sources fell by half between 1985 and 1995, undoing most of the gains made during the 1970s. Consumers purchased gas-guzzling trucks, jeeps, and minivans, which lowered the average fuel efficiency of the U.S. fleet, and energy demand resumed its upward growth (Salpukas 1997; Schaeffer 1997b, chap. 11). The return of "plenty" in the 1980s undermined environmental arguments about "scarcity" and deflected or deferred fears about the "limits to growth." Whereas economic developments (inflation) had contributed to the environmental movement's success in the 1970s, economic developments (deflation) in the 1980s undermined public support for movement policies (limits to growth) and practices (conservation).

But while deflation undermined public support for the environmental movement in the core, it contributed to the growth of the environmental movement in the periphery. Debt crisis and austerity programs in the periphery accelerated the production of energy, food, and natural resources, resulting in the conversion of small-scale subsistence agriculture to export and feed crops and the destruction of tropical forests (Schaeffer 1997b, chaps. 5–8). As a result, the food and resources that many people relied on for survival became more expensive and scarcer. Conditions of growing scarcity now provided economic support for environmental arguments in the periphery, much as they had previously done in the core.

Proponents of the world-system perspective have long argued that global processes—colonialism, world war, long periods of economic expansion and contraction, and environmental destruction—have different consequences for people, movements, and states in different regions or zones of the world. In this context, people in the core experienced the global inflation of the 1970s as a period of scarcity, while people in the periphery experienced it as a period of relative plenty. Conversely, for people in the core, deflation in the 1980s meant the return of plenty; for people in the periphery it meant the return of scarcity. Yet people in both zones responded in much the same way when they faced similar economic problems. Under conditions of scarcity, environmental movements grew in both the core (1970s) and periphery (1980s), while in times of plenty, environmental movements in both the core (1980s) and periphery (1970s) fared poorly. This kind of development is also anticipated by proponents of world-system theory, who have long argued that people around the world respond to common developments in similar ways. This suggests that the world-system perspective can usefully be applied to the study of environmental change and environmental movements, both at the global level and in different zones of the world-economy.

While economic developments contributed to the spread of environmental movements in the periphery, so too did political change. The democratization of states throughout the periphery in the 1980s made it possible for environmental movements there to organize in relative safety for the first time. And when environmental groups emerged in the periphery, they received, for the first time, real support from well-established (Greenpeace, World Wildlife Fund) and newly organized (Rainforest Action Network, Pesticide Action Network) global environmental groups in the core (Schaeffer 1985, 10–11; Wapner 1996, 82–96).

Relations between environmental groups in the core and in the periphery improved dramatically in the 1980s, not only because core groups provided important technical assistance and real economic aid, but because core groups had revised their policies toward environmental problems in the periphery. They recognized that industry and people in the core were responsible for many environmental problems in the periphery (waste trade, global warming, debt crisis and deforestation, high protein diets in the core and hunger in the periphery) and learned to approach common problems with a new awareness of their subtlety and complexity (Schaeffer 1985, 104–06, 109, 129, 138–39, 152–53).

Greenpeace, for example, had long urged a complete moratorium on whaling and an embargo on tuna caught using methods that killed dolphins. But it revised its blanket opposition to these activities after it became apparent that a whaling moratorium conflicted with the needs of indigenous, subsistence Inuit whalers in Alaska and that the tuna embargo undercut tuna fishers in Mexico, who argued that a blanket embargo helped the U.S. tuna fleet monopolize the industry (Bonanno and Constance 1994, 118–34; Bleifuss 1996, 12–13; Ostertag 1991). As Greenpeace International Director Matti Wuori explained: "We have made mistakes in the past. Some of the solutions we have advocated in the rich North simply don't work or are harmful when applied to other countries. . . . Our [tuna-dolphin] campaign had had discriminatory effects on the poorer parts of the Western Hemisphere" (Waller 1992, 6). Efforts to appreciate the social consequences of environmental policy helped environmental groups in the core and periphery develop common agendas.

Third, some of the environmental movement's theoretical expectations failed to materialize, undermining its legitimacy and damaging important political relations with key social constituencies. And some of its political tactics failed to produce the kind of results they had achieved earlier, largely because the movement failed to develop a useful theory of its relation to the state.

By the early 1980s, it became apparent that Ehrlich's theories about the consequences of rapid population growth were wholly inadequate and that events in the early 1970s had masked important social, economic, and demographic developments that prevented the Population Bomb from exploding. Ehrlich's basic theoretical expectations were wrong because the sexual revolution transformed gender relations and slowed birth rates; because the widespread adoption of green-revolution agricultural technologies increased global per capita food production and raised per capita food production even in countries like India; because it turned out that war in Bangladesh, Ethiopia, and Somalia produced famine, not famine produced war; because scientists discovered that worrisome environmental problems like global warming were more often a product of affluence and technology than a product of poverty and overpopulation; and because economic developments made the supplies of many nonrenewable resources cheaper and more "plentiful" than they had been in the 1970s (Schaeffer 1997b, chap. 10).

The collapse of this central theory had important political consequences because it removed a source of unity for different environmental constituencies. Population-control groups declined (ZPG went out of business and many of its members moved on to work on anti-immigration policy), and environmental groups dropped population as a political issue (Ferriss 1993; Dowie 1996, 160, 163–65). These developments broke the environmental movement's close association with the women's movement, just as the women's movement faced new battles over reproductive rights with the Reagan administration and its auxiliaries on the religious right. Instead of joining this fight, the environmental movement sat on the sidelines in battles over abortion, a stance that caused a breach with the women's movement (Seager 1993, 217–18).

Furthermore, tactics that had been successful in the 1970s often failed to work in the 1980s. For example, when the nuclear-power industry declined in the early 1980s, many antinuclear power groups joined with peace groups to create a movement against nuclear weapons and the arms race (Schaeffer 1989; Lewis 1984; Wheeler 1982). But grassroots activism on the Nuclear Weapons Freeze Campaign failed to accomplish the kind of political success that antinuclear-power activists had earlier achieved, and they won only a nonbinding resolution endorsing the Freeze in the House.

Grassroots tactics failed in the 1980s because authority over nuclear weapons was not dispersed or delegated to local authority, which might have been vulnerable to grassroots protest, but was centralized in the federal government where it was defended by the military–industrial complex. Not only was authority over nuclear weapons policy extremely centralized, but the rules of political engagement were different. Information about nuclear weapons was subject to stringent secrecy laws and protest could be treated as a felony, not simply a misdemeanor. As a result, the risks associated with nonviolent direct action were considerably higher in protests against nuclear weapons than they were in actions against nuclear power. The Freeze campaign never appreciated these differences, though they did try to "centralize" the movement by merging with SANE and moving their headquarters from St. Louis to Washington, D.C. (Lewis 1984, 8). When Reagan and Gorbachev signed a modest arms control treaty in 1987, SANE-Freeze quickly collapsed and the movement dispersed (*Nuclear Times,* 1985–1987).

The failure of grassroots antinuclear activism in the United States, and in Western Europe, demonstrated the movement's inability to theorize adequately its relation with the state (Prendiville 1994; Lyons 1988). There have been several obstacles to useful theorizing. One problem was that a singular relation to the state can have both beneficial and negative consequences. The movement's close association with the U.S. state in the 1970s helped pass important domestic legislation, but harmed its relations with movements and states in the periphery. Conversely, when the movement was denied access to the U.S. state, it was unable to pass domestic legislation though it was able to improve its relations with states and movements in the periphery.

A second problem was that the movement had very different relations with the state at the same time. Greenpeace, for example, conducted three important campaigns in the late 1980s, each with a different relation to the U.S. state. The marine mammals and Antarctica campaigns worked closely and cooperatively with the U.S. Commerce Department, which for a variety of reasons supported the antiwhaling moratorium in the International Whaling Commission and supported efforts to protect Antarctica. The toxics campaign, which worked closely with grassroots antitoxic groups around the country, had an adversarial relation with the EPA, though relations improved somewhat after William Reilly took over the EPA. By contrast, the nuclear-free seas campaign had an extremely hostile relation with the U.S. Navy, which became acute after Navy warships rammed the *M.V. Greenpeace* during a protest against a Trident-missile test in international waters. Because the movement had different relations with the state, which varied from one issue to another, it was difficult for it to theorize a uniform approach to state power.

A third problem was that the movement's relation to the state varied over time. As we have seen, a strategy that proved effective in one period sometimes proved ineffective in another, depending on how the state organized power and delegated its authority. Generally, grassroots tactics were effective when state power was devolved to local authority (hence the success of grassroots civil rights, antinuclear power, and antitoxics campaigns), but ineffective when state power was centrally held (hence the failure of grassroots antiwar and antinuclear-weapons campaigns). This also held true in Western Europe, where grassroots antinuclear-power and antinuclear-weapons campaigns both failed, largely because authority over nuclear power and nuclear weapons was centrally held in Western European states.

A final problem was that the movement has not paid much attention to political theorizing, largely because the publications of its member organizations are devoted to publicizing the organization's activities, not analyzing or assessing the work of other organizations or the movement as a whole. The movement's inability to theorize multiple, complex, and historically contingent relations to state power has meant that it has been unable to find a way out of its current political impasse. There has been little recognition in the current debate, which is largely conducted between the conservationist G-10 groups and the environmentalist grassroots components of the movement, that any singular strategy (lobbying or direct action) would likely fail if adopted by the movement as a whole (see Zisk 1992).

In this context, the movement would be strengthened if the G-10 shared its resources with grassroots and minority groups in the United States and abroad, much as Greenpeace International shares the resources of countries in the core with Greenpeace affiliates in the periphery; if it repaired its relations with the women's movement by renewing efforts to protect and extend reproductive rights; and if it established new relations with the labor movement, particularly on trade-related issues. These steps would help expand and consolidate the movement's social base. In economic terms, the movement should argue that while conditions of "plenty" have returned in the core, they have returned only for some, and that "scarcity" remains a problem for many in the core and for most in the periphery. This means that the movement should pay more attention to domestic and global inequalities in wealth and argue that debt, free trade, and other core economic policies adversely affect people and the environment around the world. And in political terms, the movement should probably distance itself both from the state and the Democratic Party, while recognizing that its relation to the state and political power is complex and varied.

Of course, the ongoing impasse of the environmental movement, now more than 15 years old, has not led to a collapse of the movement, to the kind of organizational failure that visited the New Left, civil rights, and peace movements in earlier decades. The environmental movement has not collapsed because it had deeper organizational roots (many conservationist groups are now a century old), because it was more diverse (with organizations ranging from the Nature Conservancy to the Mothers of East L.A.), because it was less centralized (member groups never created a central organization like Students for a Democratic Society or SANE-Freeze), and because it has been more adaptable (Dave Brower's particular kind of conservationist–environmentalism has taken many organizational forms: the Sierra Club, Environmental Policy Center, Friends of the Earth, Earth Island) (Walsh 1985; Raubner 1986). Over time, the movement has recruited new social groups, including poor and minority groups, and maintained high levels of public consciousness largely because it has made environmental education an integral part of elementary, high school, and college curriculums (Dunlap 1995, 97; Dowie 1996, 32). Because the environmental movement has not collapsed, there is good reason to hope that it will eventually find a way out of its current impasse.

NOTES

1. Dowie says that "23 federal environmental acts were signed into law" during the 1970s (1996, 33). In addition to these laws, the government created the Council on Environmental Quality and the Environmental Protection Agency, charging these and other agencies with the responsibility for implementing new laws. Lester 1995, 3; Scheffer 1991, 143–44, 150–66. The Nixon administration in 1969 also banned DDT use in the United States. Cable and Cable 1995, xvii–xxii. In 1969, President Nixon announced that the government would "give population [control] and family planning a high priority," called on other governments to take "prompt action" to slow population growth, and later lobbied the United Nations to convene its first World Population Conference in Bucharest, which was held in 1974 (Mass 1976, 63).

2. Dunlap and Mertig 1992, 2. Polls show that concern for the environment increased dramatically for a few years after Earth Day. By 1990, about 75 percent of Americans called themselves "environmentalists," and 45 percent believed that "protecting the environment is so important that requirements and standards cannot be too high and continuing environmental improvements must be made regardless of cost" (Dowie 1996, 4; Dunlap 1992, 113; Dunlap 1995, 99; Kamieniecki 1995, 162). Voluntary recycling, for example, has become a common practice for 80 percent of Americans and a large majority claims to have made "changes in [their] day-to-day behavior because of [their] concern about the environment." Participation in recycling has grown from 33 percent in 1972 to 80 percent in 1990 (Dunlap 1995, 101–02).

3. Membership in the major environmental organizations grew from 819,000 in 1969 to nearly 2 million in 1983, and then to more than 3 million in 1990, while tens of thousands more participated in local grassroots groups, and the collective income of the major groups grew to $217.3 million by 1990 (Ingram, Colnic and Mann 1995, 122; Mitchell 1991, 96; Mitchell, Mertig, and Dunlap 1992, 13, 15, 18; Szasz 1994, 72).

4. One is reminded of Henry Kissinger's conversation with Mao Tse-tung. Kissinger asked Mao what he though had been the results of the French Revolution. And Mao replied: "It's too early to tell" (Shabecoff 1996, 171).

5. McCormick 1989, 15; Schnaiberg and Gould 1994, 148–49.

6. Glenn notes that "transatlantic protests against corporal punishment occurred concurrently with Anglo-American campaigns against slavery, capital punishment, dueling, war, and cruelty to animals" (1983, 409).

7. Glenn 1983, 409; McCormick 1989, 4–5. At first, they were particularly concerned with the flogging of horses, children, slaves, and sailors. Later they became concerned about more general cruelties.

8. Nelson, a war veteran, was then 53 years old; Hayes was an antiwar activist and law student at Harvard. Schwab 1994, 29–30, 39–43.

9. For the young, the baby boom meant huge, impersonal universities; for the old, it meant the spread of suburban housing developments and freeways across the landscape.

10. The "old money" supplied by the Ford Foundation played an important role in knitting the two groups together. Mitchell 1991, 89; Ingram, Colnic, and Mann 1995, 118.

11. Mitchell 1980, 218. Other writers argued that resource scarcity "might actually be turned into a grand opportunity to build a more humane postindustrial society" (Buttel and Larson 1980, 344). But James O'Connor differed, arguing that "ecological economists unwittingly strengthened the hands of the neoclassical economists' claim that consumption was too high by pointing to shortages of natural resources and dangerous levels of pollution." (1987, 33–34).

12. A common perspective was advanced by Denis Hayes, who argued: "I suspect that politicians and businessmen who are jumping on the environmental bandwagon don't have the slightest idea what they are getting into" (Schwab 1994, 11).

13. Nixon 1972, 167. This, incidently, is a theme that David Brower has recently revived, arguing not only for conservation but "restoration."

14. International environmental groups with roots in the United States became among the largest environmental groups in many European countries (Dalton 1994, 41, 54, 58, 89–90, 95; Lyons 1988, 26; Prendiville 1994, 7, 86; Seager 1993, 177).

15. Warner and David Kreger, "Friends All Over Europe," *Not Man Apart,* February–March, 1984, 17. LaLond would later join the Mitterand government as minister of the environment in 1988 (Dalton, 1994, 216).

16. Switzer 1994, 42. The U.S. environmental-movement's criticism of the French nuclear-weapons testing program in the Pacific antagonized the French and undermined public support for environmental issues in France. Dalton 1994, 90.

17. *Silent Spring, The Fate of the Earth,* and *The End of Nature* all appeared first in the *New Yorker*. Its role as the unofficial theoretical journal of the environmental movement has been unappreciated.

18. Rachel Carson set the tone in *Silent Spring,* which began with this warning from Albert Schweitzer: "Man has lost the capacity to foresee and forstall. He will end by destroying the Earth."

19. Hal Lindsay's fundamentalist tome, *The Late Great Planet Earth,* which appeared in 1970, sold nine million copies by 1978. Boyer 1992, 5.

20. Health officials estimate that 750,000 people are annually poisoned by pesticides around the world, and as many as 14,000 die. Matthiessen 1992; Briggs 1990; Schaeffer 1985, 11.

21. "Blues for Most Green," *Greenview* 1994. Greenpeace lost more than one million donors between 1990 and 1994. "Environmental Movement Fights for Lost Momentum," *Oakland Tribune,* September 22, 1994; Dowie 1996, 41, 46, 70, 175, 193–94.

22. Christrup and Schaeffer 1990; Bullard 1994, 6. They fought not only for themselves but for others. Although they are often described as NIMBYs (Not In My Back Yard), they should more accurately be described as NIABYs (Not In Anyone's Back Yard), a term I coined, because, as one activist explained: "No one wants to send leukemia to another community. The American people just aren't that way." Christrup and Schaeffer 1990, 18.

23. Kreger and Warner 1984, 10. The debate over free trade also split the movement. Dowie 1996, 188; Schaeffer 1995, 270.

24. Schaeffer 1984a, 11; Schneider 1995. It was also the case that foundations withdrew their support for environmental issues and began funding other, more "pressing" issues. Dowie 1996, 176.

REFERENCES

Andrews, Richard N. L. 1980. Class Politics or Democratic Reform: Environmentalism and American Political Institutions. *Natural Resources Journal* 20(2): 227, 231.

Bandarage, Asoka. 1994. A New Malthusianism. *Peace Review* 6(3): 294.

Bleifuss, Joel. 1996. The Great Dolphin Divide. *In These Times,* August 5, 12–13.

Bloch, Ben, and Harold Lyons. 1993. *Apocalypse Not: Science, Economics and Environmentalism.* Washington, DC: Cato Institute.

"Blues for Most Green," 1994. *Greenview* 4(119): 1–5.

Bonanno, Alessandro, and Douglas Constance, 1994. The Global Agri-Food Sector and the Case of the Tuna Industry: Global Regulation and Perspective for Development. *International Journal of Sociology of Agriculture and Food* 4: 118–34.

Boyer, Paul. 1992. *When Time Shall Be No More: Prophecy Belief in Modern American Culture.* Cambridge, MA: Belknap Press.

Branch, Taylor. 1988. *Parting the Waters: America in the King Years, 1954–63.* New York: Simon & Schuster.

Briggs, Shirley A. 1990. Silent Spring: The View from 1990. *The Ecologist* 20(2): 6–8.

Bullard, Robert D. 1994. *Dumping in Dixie: Race, Class, and Environmental Quality.* Boulder, CO: Westview Press.

Buttel, Frederick H. 1995. Rethinking International Environmental Policy in the Late Twentieth Century. In *Environmental Justice: Issues, Policies and Solutions,* ed. Bunyan Bryant, 187–98. Washington, DC: Island Press.

Buttel, Frederick H., and Oscar W. Larson III. 1980. Wither Environmentalism? The Future Political Path of the Environmental Movement. *Natural Resources Journal* 20(2): 326.

Cable, Sherry, and Charles Cable. 1995. *Environmental Problems: Grassroots Solutions: The Politics of Grassroots Environmental Conflict.* New York: St. Martin's Press.

Caldwell, Lynton K. 1992. Globalizing Environmentalism: Threshold of a New Phase in International Relations. In *American Environmentalism: The U.S. Environmental Movement, 1970–1990,* ed. R. E. Dunlap and A. G. Mertig, 63–76. Philadelphia: Taylor & Francis.

Christrup, Judy, and Robert Schaeffer. 1990. Not in Anyone's Backyard. *Greenpeace* (January–February): 14–19.

Costain, W. Douglas, and James P. Lester. 1995. The Evolution of Environmentalism. In *Environmental Politics and Policy: Theories and Evidence,* ed. J. P. Lester, 15–38. Durham, NC: Duke University Press.

Dalton, Russell J. 1994. *The Green Rainbow: Environmental Groups in Western Europe.* New Haven, CT: Yale University Press.

Devall, Bill. 1980. The Deep Ecology Movement. *Natural Resources Journal* 20(2): 303, 319–20.

Devall, Bill. 1992. Deep Ecology and Radical Environmentalism. In *American Environmentalism: The U.S. Environmental Movement, 1970–1990,* ed. R. E. Dunlap and A. G. Mertig, 51–62. Philadelphia: Taylor & Francis.

Devall, Bill, and George Sessions. 1985. *Deep Ecology.* Salt Lake City: Gibbs M. Smith.

DeWitt, John. 1994. *Civic Environmentalism: Alternatives to Regulation in States and Communities.* Washington, DC: Congressional Quarterly Press.

Dowie, Mark. 1996. *Losing Ground: American Environmentalism at the Close of the Twentieth Century.* Cambridge, MA: MIT Press.

Dunlap, Riley E. 1992. Trends in Public Opinion Toward Environmental Issues, 1965–1990. In *American Environmentalism: The U.S. Environmental Movement, 1970–1990,* ed. R. E. Dunlap and A. G. Mertig, 89–116. Philadelphia: Taylor & Francis.

Dunlap, Riley E. 1995. Public Opinion. In *Environmental Politics and Policy: Theories and Evidence,* ed. J. P. Lester, 63–114. Durham, NC: Duke University Press.

Dunlap, Riley E., and Angela G. Mertig, eds. 1992. *American Environmentalism: The U.S. Environmental Movement, 1970–1990.* Philadelphia: Taylor & Francis.

Dunlap, Riley E., and Angela G. Mertig. 1992. The Evolution of the U.S. Environmental Movement from 1970 to 1990: An Overview. In *American Environmentalism: The U.S. Environmental Movement, 1970–1990,* ed. R. E. Dunlap and A. G. Mertig, 1–10. Philadelphia: Taylor & Francis.

Easterbrook, Gregg. 1990. Everything You Know About the Environment Is Wrong. *New Republic* (April 30): 14–27.

Ehrlich, Paul. 1968. *The Population Bomb.* New York: Ballentine Books.

Ferriss, Susan. 1993. FAIR: Mounting Campaign to Keep Immigrants Out. *San Francisco Examiner,* December 12.

Freudenberg, Nicholas, and Carol Steinsapir. 1992. Not in Our Backyards: The Grassroots Environmental Movement. In *American Environmentalism: The U.S. Environmental Movement, 1970–1990,* ed. R. E. Dunlap and A. G. Mertig, 27–38. Philadelphia: Taylor & Francis.

Gilliam, Harold. 1995. Does Anyone Care about the Environment Anymore? *San Francisco Chronicle,* March 5.

Gitlin, Todd. 1987. *The Sixties: Years of Hope, Days of Rage.* New York: Bantam.

Glenn, Myra C. 1983. The Naval Reform Campaign against Flogging: A Case Study in Changing Attitudes Toward Corporal Punishment, 1830–1850. *American Quarterly* 35(4): 409.

Grossman, Karl. 1994. The People of Color Environmental Summit. In *Unequal Protection: Environmental Justice and Communities of Color,* ed. Robert D. Bullard, 272–77. San Francisco: Sierra Club Books.

Ingram, Helen M., David H. Colnic, and Dean E. Mann. 1995. Interest Groups and Environmental Policy. In *Environmental Politics and Policy: Theories and Evidence*, ed. J. P. Lester, 115–45. Durham, NC: Duke University Press.

Kaiser, Sandra. 1983. Cheap Oil: Such a Bargain? *Not Man Apart* (April), 2.

Kamieniecki, Sheldon. 1995. Political Parties and Environmental Policy. In *Environmental Politics and Policy: Theories and Evidence,* ed. J. P. Lester, 146–67. Durham, NC: Duke University Press.

Kaufman, Wallace. 1994. *No Turning Back: Dismantling the Fantasies of Environmental Thinking.* New York: Basic Books.

Komanoff Energy Associates. 1992. *Fiscal Fission: The Economic Failure of Nuclear Power.* December.

Kreger, David, and Gale Warner. 1984. The Reagan Presidency. *Not Man Apart* (September): 1–16.

Kriz, Margaret. 1989. Ahead of the Feds. *National Journal* (December 9): 2991.

Lester, James P. 1995. Federalism and State Environmental Policy. In *Environmental Politics and Policy: Theories and Evidence,* ed. J. P. Lester, 39–60. Durham, NC: Duke University Press.

Lester, James P., ed. 1995. *Environmental Politics and Policy: Theories and Evidence.* Durham, NC: Duke University Press.

Lewis, David. 1984. Tough Choices for the Freeze. *Not Man Apart* (October): 8.

Lewis, Martin W. 1992. *Green Delusions: An Environmentalist Critique of Radical Environmentalism.* Durham, NC: Duke University Press.

Lilienfeld, Robert M., and William L. Rathje. 1993. Six Enviro-Myths. *New York Times,* January 21.

Luker, Kristin. 1984. *Abortion and the Politics of Motherhood.* Berkeley: University of California Press.

Lyons, Matthew Nemiroff. 1988. *The "Grassroots" Network: Radical Nonviolence in the Federal Republic of Germany, 1972–1985.* Ithaca, NY: Center for International Studies, Cornell University.

Mass, Bonnie. 1976. *Population Target: The Political Economy of Population Control in Latin America.* Toronto: Women's Press.

Matthiessen, Constance. 1992. The Day the Poison Stopped Working. *Mother Jones* (March–April): 16.

McClosky, Michael. 1992. Twenty Years of Change in the Environmental Movement: An Insider's View. In *American Environmentalism: The U.S. Environmental Movement, 1970–1990,* ed. R. E. Dunlap and A. G. Mertig, 77–88. Philadelphia: Taylor & Francis.

McCormick, John. 1989. *Reclaiming Paradise: The Global Environmental Movement.* Bloomington: Indiana University Press.

Milbrath, Lester W. 1984. *Environmentalists: Vanguard for a New Society.* Albany: SUNY Press.

Mitchell, Robert Cameron. 1991. From Conservation to Environmental Movement: The Development of the Modern Environmental Lobbies. In *Government and Environmental Politics: Essays on Historical Developments Since World War Two,* ed. Michael J. Lacy, 81–14. Washington, DC: Woodrow Wilson Center Press.

Mitchell, Robert Cameron. 1980. Introduction. *Natural Resources Journal* 20(2): 217–19.

Mitchell, Robert Cameron, Angela G. Mertig, and Riley E. Dunlap. 1992. Twenty Years of Environmental Mobilization: Trends among National Environmental Organizations. In *American Environmentalism: The U.S. Environmental Movement, 1970–1990,* ed. R. E. Dunlap and A. G. Mertig, 11–26. Philadelphia: Taylor & Francis.

New York Times. 1971. August 16.

Nixon, Richard M. 1972. Environmental Quality. In *Politics, Policy and Natural Resources,* ed. Dennis L. Thompson, 156–67. New York: Free Press.

O'Connor, James. 1987. *The Meaning of Crisis: A Theoretical Introduction.* Oxford, UK: Basil Blackwell.

O'Rourke, P. J. 1992. The Greenhouse Affect. In *The Rolling Stone Environmental Reader,* ed. Jann S. Wenner, 15–24. Washington, DC: Island Press.

Ostertag, Bob. 1991. Greenpeace Takes Over the World. *Mother Jones* (March–April): 32.

Pope, Carl. 1972. Population. In *Nixon and the Environment: The Politics of Devastation,* ed. James Rathlesberger, 163–79. New York: Village Voice Books.

Prendiville, Brendan. 1994. *Environmental Politics in France.* Boulder, CO: Westview Press.

Pulido, Laura. 1996. *Environmentalism and Economic Justice: Two Chicano Struggles in the Southwest.* Tucson: University of Arizona Press.

Rathlesberger, James. 1972. *Nixon and the Environment: The Politics of Devastation.* New York: Village Voice Books.

Raubner, Paul. 1986. With Friends Like These. *Mother Jones* (November): 35.

Resenbaum, Walter A. 1995. The Bureaucracy and Environmental Policy. In *Environmental Politics and Policy: Theories and Evidence,* ed. J. P. Lester, 206–41. Durham, NC: Duke University Press.

Roszak, Theodore. 1992. Green Guilt and Ecological Overload. *New York Times,* June 9.

Rushefsky, Mark E. 1995. Elites and Environmental Policy. In *Environmental Politics and Policy: Theories and Evidence,* ed. J. P. Lester, 275–99. Durham, NC: Duke University Press.

Salpukas, Agis. 1997. Green Power Wanes, But Not at the Grassroots. *New York Times,* March 9.

Sardar, Ziauddin. 1981. Malaysia Silences FOE and Other Groups. *Not Man Apart* (June): 6.

Schaeffer, Robert K. 1983. Green Pitch for Peace. *Not Man Apart* (November): 12.

Schaeffer, Robert K. 1984a. Green Envy. *Not Man Apart* (September): 10–11.

Schaeffer, Robert K. 1984b. California Upholds Aerial Spraying Ban. *Not Man Apart* (September): 5.

Schaeffer, Robert K. 1985. Bhopal Spurs Pesticide Reform Movement. *Not Man Apart* (July–August): 10–11.

Schaeffer, Robert K. 1989. Anti-Nuclear Families. *Nuclear Times* (January–February): 23–25.

Schaeffer, Robert K. 1995. Free Trade Agreements: Their Impact on Agriculture and the Environment. In *Food and Agrarian Orders in the World-Economy,* ed. Philip McMichael, 255–75. Westport, CT: Praeger.

Schaeffer, Robert K. 1997a. *Power to the People: Democratization around the World.* Boulder, CO: Westview Press.

Schaeffer, Robert K. 1997b. *Understanding Globalization: The Social Consequences of Political, Economic and Environmental Change.* Lanham, MD: Rowman & Littlefield.

Scheffer, Victor B. 1991. *The Shaping of Environmentalism in America.* Seattle: University of Washington Press.

Schnaiberg, Allan, and Kenneth Alan Gould. 1994. *Environment and Security: The Enduring Conflict.* New York: St. Martin's Press.

Schneider, Keith. 1992. Thwarted Environmentalists Find U.S. Courts Are Citadels No More. *New York Times,* March 22.

Schneider, Keith. 1994. For the Environment, Compassion Fatigue. *New York Times,* November 6.

Schneider, Keith, 1995. Big Environment Hits a Recession. *New York Times,* January 1.

Schwab, Jim. 1994. *Deeper Shades of Green: The Rise of Blue-Collar and Minority Environmentalism in America.* San Francisco: Sierra Club Books.

Seager, Joni. 1993. *Earth Follies: Coming to Feminist Terms with the Global Environmental Crisis.* New York: Routledge.

Shabecoff, Philip. 1996. *A New Name for Peace: International Environmentalism, Sustainable Development and Democracy.* Hanover, NH: University Press of New England.

Shabecoff, Philip. 1989. The Environment as a Local Jurisdiction. *New York Times,* January 22.

Stevens, William K. 1993. What Really Threatens the Environment? *New York Times,* January 21.

Switzer, Jacqueline Vaugh. 1994. *Environmental Politics: Domestic and Global Dimensions.* New York: St. Martin's Press.

Szasz, Andrew. 1994. *EcoPopulism: Toxic Waste and the Movement for Environmental Justice.* Minneapolis: University of Minnesota Press.

Teitelbawm, Michael S. 1992/93. The Population Threat. *Foreign Affairs* (Winter): 63–78.

Teodori, Massimo. 1969. *The New Left: A Documentary History.* New York: Bobbs-Merrill.

Tierney, John. 1996. Recycling Is Garbage. *New York Times Magazine,* June 30.

Tucker, Richard P. 1981. India Gets Serious about Conservation. *Not Man Apart* (May), 10–11.

Tucker, William. 1982. *Progress and Privilege: America in the Age of Environmentalism.* Garden City, NY: Anchor Press.

Turner, Tom. 1990. *Wild by Law: The Sierra Club Legal Defense Fund and the Places It Has Saved.* San Francisco: Sierra Club Legal Defense Fund and Sierra Club Books.

Waller, Mark. 1992. Behind New Leader, Greenpeace Reexamines Itself. *In These Times* (February 12–18), 6.

Walsh, Joan. 1985. At the Crossroads, Friends of the Earth Confront Difficult Choices. *In These Times* (April 3–9).

Wapner, Paul. 1996. *Environmental Activism and World Civic Politics*. Albany: SUNY Press.

Warner, Gale, and David Kreger. 1984. Friends All Over Europe. *Not Man Apart* (February–March): 17.

Wheeler, Steve. 1982. Support Builds for a Nuclear Freeze. *Not Man Apart* (May), 18.

Zisk, Betty H. 1992. *The Politics of Transformation: Local Activism in the Peace and Environmental Movements*. Westport, CT: Praeger.

11

Globalization, Democratization, and the Environment in the New South Africa: Social Movements, Corporations, and the State in South Durban

Christine Root & David Wiley, with Sven Peek

INTRODUCTION

In several semiperipheral countries, environmental movements only recently have begun to address the environmental costs of expanded industrial production. South Africa, a semiperipheral state, is undergoing a dramatic transition to democracy after nearly 50 years of apartheid rule. This transition has created new political space for environmental social movements to address its accumulated environmental problems and woefully inadequate environmental regulation.

South Africa faces a vast legacy of environmental problems from the apartheid era—the degraded land in former homelands, the over-exploited marginal drylands, the enormous wastes of the mining industry, air pollution and global warming gases produced from coal-based energy production, and the lack of control of toxics and chemical pollution from industry.

A few local struggles against environmental racism occurred during the antiapartheid mobilization of the 1980s and early 1990s. These struggles, and the broader antisystemic mobilization against apartheid, created the groundwork for the current environmental movement.

We have focused on the sector of the South African environmental movement waging struggles against urban industrial air pollution. We describe and analyze two local environmental forums dealing with this issue in the southern industrial basin of Durban, from among seven forums there that we studied during more than two years of research.[1]

We sought to understand how the impacts of increasing integration into the world-system of South African industries, government, and the environmental movement as well as South Africa's new political culture of democratic consultation either enhanced or constrained the power of the locally based environmental movement in its struggles against industrial pollution. More specifically, we assessed how the community-based environmental movement (CBEM) generated its political power, how it exercised power in the politics of interactions with companies and the local and national government, and how it utilized resources and linkages with environmental movements abroad. The CBEM comprised community-based civic organizations (CBOs), community-based environmental organizations, and local chapters of environmental nongovernmental organizations (NGOs).[2]

THEORETICAL PERSPECTIVES

Much of the literature on environmental disputes fails to base its analysis on the unequal power relations between corporations and social movements. As Modavi points out about many studies in the United States, "disputes are transformed from conflicts that are often between the grass-roots and industry groups to that of paradigm clashes of groups with equal power by differing ontological realities" (1996, 303).

A class-based analysis that recognizes unequal power relations posits that creating new structures of participation will not necessarily level the playing field or enhance power of the working class or social movements. Focusing on corporate and union interactions in European countries taking the corporatist path, Offe (1985) posits that, because of this imbalance of power, corporations and labor unions generate and use their power in different ways. This theory is useful to understanding interactions between corporations and environmental social movements as well.

Offe and several other theorists posit that corporations also have greater access to the state than do unions and social movements. The state faces contradictory tasks in relation to different classes. "On the one hand, the state must sustain the process of accumulation and the private appropriation of resources; on the other hand, it must preserve belief in itself as the impartial arbiter of class interest, thereby legitimating its power" (LeRoux 1996, 276). Even relatively progressive states with which the working class has some influence face the dilemma of needing corporate investment in order to provide a surplus for public social spending for the poor.

When many of these corporations are transnational in character—as is predominant in South Africa and other semiperipheral countries—organizations engaged in local environmental struggles face a dilemma. According to Gould, Schnaiberg, and Weinberg (1996):

> Current trends indicate that, increasingly, economic flows, waste flows, cultural flows, and depletion patterns are international and transnational in scope. In contrast, environmental political mobilization and state-sponsored public participation schemes are organized more frequently at the community level. . . . The socioeconomic causes of environmental problems are therefore organized at a different level than are the social responses to these problems. The question this paradox poses for the modern environmental movement is: can local-level mobilization effectively constrain transnational-level production?

They posit that locally based environmental movements have very limited power in relation to these transnational corporations and that those that are successful generally achieve only small gains.

Nevertheless, because they are strongly motivated to take action on issues that directly affect them and their families, it is not likely that people will abandon these struggles. In these conflicts, locally based environmental movements can increase their power by building local, national, and transnational alliances. Such alliances sometimes bring some improvement in the lives of the communities that initiate them and even affect local and national policy. Indeed, the transnational scope of corporations may even create the possibility of transnational alliances by their opponents.

THE IMPACTS OF GLOBALIZATION ON THE ENVIRONMENTAL STRUGGLE IN SOUTH AFRICA

Now that apartheid has ended after a period of economic sanctions and divestment by some transnational corporations, South Africa is being integrated into the world-system at an accelerated pace. With the extension and institutionalization of the capitalist world-system and the demise of the Soviet Union and East European socialist governments, South Africa has neither working models, international solidarity, nor support for pursuing a more progressive national economic course.

We consider four consequences of South Africa's accelerated integration into the world-system for the environmental struggles there. First, we explore impacts of global integration in two arenas—South Africa's environmental norms and regulations and the expanded linkages with environmental movements abroad. Then we turn to two broader ways in which the world-system affects policy about the environment in South Africa: pressures for democratization (which, of course, was at the core of the antiapartheid struggle and therefore is not exclusively or even primarily a global dynamic), and for neoliberal economic reform.

Global Environmental Standards and Environmental Linkages

New international standards of corporate environmental management and reporting are being applied to the production of South African goods traded abroad, including European environmental regulations on imported goods, the voluntary standards of the International Organization for Standardization (ISO), and environmental standards negotiated by the World Trade Organization. South African corporate environmental management also is influenced by the ubiquitous norms of environmental impact assessment (EIA), and the government has mandated EIAs for new investments beginning in April 1998. The United Nations Conference on the Environment and Development (UNCED) in Rio de Janeiro also has shaped the rhetoric of government environmental policy.

Several Western industrial countries have developed assistance programs to impact directly on South Africa's environmental policy. These include a Danish-funded program to formulate an integrated waste management and pollution strategy as well as the United States–South Africa Binational Commission's exchange visits and projects concerning nature conservation, environmental management, pollution control, water conservation, and other areas of environmental policy.

During the period of transition from apartheid in 1993–1995, the Canadian International Development Research Center worked with the ANC, COSATU, SANCO, and the South African Communist Party to develop recommendations to the new government for a more rigorous environmental policy (Whyte 1995). Article 24 of the new Constitution that guarantees all South Africans the right not to be injured by their environment is an important evidence that, at one level, South Africa is taking seriously the human consequences of environmental degradation.[3]

There are contradictory global influences, however. Many Western governments are reducing their own environmental regulations or their resources and political will to enforce them. Numerous studies of environmental policy in the United States have pointed to the withdrawal of financial resources needed to implement legislated environmental standards and the frequently pro-corporate bias of environmental regulatory officials.[4] These tendencies abroad can be used to justify weak regulation of the environment in South Africa, especially taken together with broader pressures to decrease government regulation and spending that we discuss later.

A second global impact on South Africa is the expanded linkages of the environmental movement with those in other countries. A "green tradition" in South Africa is maintained by many organizations, including the Wildlife Society and the International Union for Conservation of Nature, which are deeply linked internationally. Industrial pollution has spawned an active South African antitoxic movement that often collaborates with movements abroad, exchanging information and sharing strategies and campaigns.

The Impacts of Democratization in South Africa

The third global impact is the Western push for democratization that has been a strong rhetorical element of the foreign policy of the United States and other Western governments with the IMF and the World Bank in the post-Cold War era. Usually, this conception of democracy has emphasized multiparty democracy and has been closely linked to pressure for private enterprise, free access to markets, and structural adjustment.

In South Africa, the drive for democracy is deeply rooted in at least 85 years of anticolonial and antiapartheid struggle. The new structures of universal suffrage, multiparty representation in Parliament, and a Government of National Unity (GNU) guaranteeing seats in the Cabinet for minority parties were negotiated between the ANC and the former National Party government preceding the first universal suffrage election in April 1994. Western governments and the World Bank have been eager to influence the shape of newly configured municipal and provincial governments, to recommend policies toward social spending, and to help train newly elected local representatives and new bureaucrats. From within South Africa, the government born out of a resistance movement faces pressures from persistent though weakened elements of that movement for effective democratic participation in order to implement the goals for which they struggled.

The concept of local democracy or local participation—as distinct from purely electoral democracy—has many different meanings and emphases in various international settings. "Local democracy" does not necessarily mean a strengthened role for working or poor people. For example, recommendations for municipal governments to promote sustainable development from the UNCED Conference in Rio de Janeiro in June 1992 urged consultation with all "local, civic, community, business, and industrial organizations" (UNCED 1992). This approach implicitly assumes a political parity among these highly diverse actors. On the other hand, some grassroots and nongovernment organizations in the South have developed models of "people-centered development" that explicitly aim to enhance the power of those who are most affected by poverty and environmental degradation.[5]

In South Africa today, democracy, local participation, and transparency are watchwords of the new political culture. As President Mandela has stated:

> One of the hallmarks of the democracy which South Africans are creating is the spirit of partnership which permeates our society. In policy matters in particular, the readiness of government to consult, and of civil society to share responsibility, are amongst the most striking ways in which we have broken with the past. (Mandela 1995)

The concept of democracy is enshrined in the Reconstruction and Development Programme (RDP), which calls for "[fostering] a wide range of institutions of participatory democracy in partnership with civil society. . . ." The ANC adopted the RDP as its principle policy statement in the 1994 election, and a revised version was adopted by the GNU. Originally, the RDP was conceived by the Congress of South African Trade Unions (COSATU) as a mandate for the ANC to pursue economic redistribution and to meet basic needs of the previously disenfranchised majority. The progressive movement sought to institutionalize structures that would give unions and CBOs access to local and national government decision-making on matters of policy and allocation of public resources.[6]

An explosion of new forums and consultative committees has occurred to obtain input from civil society. The National Economic Development and Labour Council (NEDLAC) is the preeminent corporatist structure designed to consult and develop consensus on economic and labor policy among business, national government, and unions. At the local community level, development forums have been created across the country to engage community representatives in addressing local infrastructural needs. Many of these forums focusing on housing and residential services attempted to construct binding "social compacts" among local community forums, government agencies, and funders. These compacts have been promoted both by supporters of local participation in building sustainable development and by large financial institutions seeking to obtain communities' commitments to development projects as a precondition for their funding (Mayekiso 1996, 243).

The Consultative National Environmental Policy Process (CONNEPP) was initiated in 1995 to engage the public in recommending revisions in national environmental policy. In 1997, for more permanent structures of environmental consultation, national, provincial, and municipal governments are being mandated to form Environmental Advisory Forums. In addition, South Africa's three largest cities (including Durban) have adopted LA 21 from the 1992 Brazil UNCED conference. The Durban municipal government envisions LA 21 as a "people-centered process that provides meaningful opportunities to develop partnerships for change" (ICLEI 1995).

Generally, these environmental forums are explicitly advisory and are accorded lower priority and influence than forums on economic issues. Some national guidelines for consultation have been drafted, particularly by the Department of Water Affairs and Forestry concerning waste landfills and water policy and the EIA guidelines more recently announced by DEAT. But still, few stakeholders have experience with participatory structures designed to address specific environmental problems involving industry such as those we studied in South Durban.

International Pressures for Neoliberal Economic Policies

A fourth global impact is the pressure on South Africa to adopt neoliberal economic policies and to encourage foreign investment. Some people in the progressive social movement in South Africa believed that President Mandela's heroic stature at the time of his election gave his new government a unique, brief moment of opportunity to resist these pressures from the IMF and World Bank. However, the government has responded to the pressure from the world-system without and the corporate sector within by implementing structural adjustment and removing trade barriers even more quickly than the IMF had suggested (Padayachee 1996, 369).

In June 1996, the government unveiled its macroeconomic "Growth, Employment, and Redistribution" (GEAR) strategy. This economic strategy seeks growth, job creation (that it assumes will accompany growth), and improvement of South Africa's competitiveness in the context of "the increasing globalisation of the world's economies" (Ramaphosa 1996). As President Mandela said in a speech about GEAR, a successful economic strategy,

> requires rapid economic growth; . . . investments that create jobs; . . . that we spend within our means as government and spend mainly in socially productive sectors; . . . that we take measures that will prevent galloping price rises; it requires that we acknowledge the realities of the world in which we live. (Mandela 1996)

In order to show fiscal restraint and thus engender investor confidence, GEAR seeks to reduce the government deficit as a proportion of the Gross Domestic Product. The GEAR strategy also includes reducing spending on social services (in order to meet the deficit target) and liberalizing the financial, labor, and trade markets.

Seeking to build development on GEAR, the ANC government also is courting foreign investors. To this end, it is heavily promoting large-scale "fast-track" investment—particularly by transnational firms—in Spatial Development Initiatives (SDIs), where regulatory "obstacles" will be reduced and investors will receive tax breaks and accelerated depreciation.

The GEAR policy has driven a large wedge between COSATU and the ANC-led government. At its Sixth Annual Congress in September 1997, COSATU roundly criticized the government for adopting GEAR without consultation, and COSATU President John Gomomo dubbed GEAR "the reverse gear of our society." Union officials noted that the premise that increased investment will automatically yield a growth in jobs already has proven false, pointing to 1996 when manufacturing production grew but 33,000 jobs were lost in that sector. The COSATU Congress adopted a resolution rejecting GEAR as unacceptable, but stopped short of calling for its abandonment in order to avert a confrontation with the ANC, when attention is focused already on the 1999 elections. It also adopted a resolution for mobilizing workers to resist the trends of globalization that undermine the sovereignty of state, decrease social security for workers, and result in rampant financial speculation. GEAR has created conflict even within the ANC. Like COSATU, ANC party structures were not involved in formulating the GEAR policy. An ANC audit of achievements of the RDP blamed the GEAR for demoralizing local party cadres by "eclipsing the ANC's RDP goals and crushing its spirit" (Edmunds and Rossouw 1997).

The four impacts of South Africa's global reintegration that we have discussed have contradictory and ambiguous effects on the South African environmental movement. The diverse global linkages that are being expanded by the South African state, as well as by corporations and the environmental movements, affect the resources and power of all of these actors, and their consequences can be assessed only by analyzing the outcomes of specific environmental struggles.

TWO STUDIES OF NEGOTIATING ON INDUSTRIAL POLLUTION

The locus of our research is South Durban, the site of the country's second-largest concentration of petroleum and petrochemical industries and where Black[7] communities under apartheid were placed immediately adjacent to these industries. South Durban adjoins the largest port on the continent, which is a key hub for South Africa's international commerce.

The Problem of Air Pollution in South Durban

South Durban has one of the highest ambient levels of sulfur dioxide (SO_2) in the country (Diab and Preston-Whyte 1996). A 1994 study in Merebank, an Indian residential community located immediately adjacent to the ENGEN refinery in South Durban, found that primary school students had in excess of three times more chronic coughs, chest congestion, and persistent wheezing than similar pupils a few kilometers away (Kistnasamy 1994).

South Africa's Atmospheric Pollution Prevention Act 45 of 1965 is widely regarded as inadequate and outdated.[8] The administration of air-pollution control is highly subjective and not sufficiently transparent, jurisdiction is fragmented, fines are trivial and do not act as a deterrent, and "it is relatively easy for any industrial concern to deviate from the conditions of its registration certificate with impunity, simply because the existing system of inspection cannot cope" (Petrie, Burns, & Bray 1992, 454).

Corporations in South Durban

Durban's history as a semiperipheral outpost of the global empire has shaped the juxtaposition of polluting industry and communities and their struggles over pollution. The policy of labor control employed when this industrial area was configured was used to separate the Indian, Coloured, and African residential communities in the basin and regulate their resources and capacities.

Beginning in the 1930s, South Durban was designed by a series of Durban City plans and ordinances to provide a center for industrial development near the harbor and housing for workers for these industries (Scott 1994). The Group Areas Act of 1950, a key building-block of the apartheid system, provided both the legal underpinnings and financial resources to carry out the segregation and formalization of these communities.

Similarly, the industries in South Durban are outgrowths of South Africa's semiperipheral position in the capitalist world-system. Some of the investment here is in processing of raw materials, such as the Mondi paper mill and the Tongaat-Hulett sugar refinery. SAPREF (owned by British Petroleum and Shell) and ENGEN (controlled by Malaysian Petronas as of 1996) refine imported oil, providing about 60 percent of the nation's petroleum. Many of the larger industries in South Durban, such as these two oil refineries and the nearby Waste-tech hazardous waste landfill, were built by foreign corporations. Others, such as the Mondi paper mill and the AECI chemicals conglomerate, are majority owned by South African-based entities. They are among the small number of large, diversified, and interconnected corporations that dominate the South African economy and, in many cases, invest heavily abroad as well.

Because it is one of the two largest sources of SO_2 pollution in South Durban and it is adjacent to two residential communities, the ENGEN refinery epitomizes for many people the problems of industry's impact on South Durban's Black communities. Mobil Oil Corporation built this first full-scale refinery in South Africa in 1954. During the liberation struggle, the apartheid regime protected and supported the strategically important oil industry, and the refineries were designated as highly secret "National Key Points."[9] According to the refinery management, the Official Secrets Act "prevented us from dealing at any level with the public about the business" (ENGEN manager, personal correspondence, 1996). The cement wall around the refinery topped by barbed wire and punctuated by guard towers attests to the fortress mentality that existed in this entire zone. Indeed, the refinery came under serious bazooka grenade fire from ANC insurgents during the struggle.

In 1989, as part of "the great trek of the multinationals" (*Southern African Economist* 1988, 14) caused by the international sanctions and divestment movements, Mobil Oil sold the refinery to GENCOR, a giant South African holding company. Although GENCOR management was largely Afrikaner, the CEO and Managing Director of this new ENGEN asset was Australian Rob Angel, who had been the CEO of Mobil's South African operations and a long-time Mobil manager in Asia.

Communities in South Durban

The neighborhoods located immediately adjacent to industry often are described by residents as "islands surrounded by a sea of polluting industries" (Chetty and Peek 1995). Three communities near the ENGEN refinery participated in forums about its pollution. Merebank is an Indian community of mixed class composition with primarily small, privately owned homes. It has a long history of protest against the expropriation and rezoning of land for industry. In the early 1990s, the Merebank Residents Association (MRA) expanded its environmental activism, seeking information, negotiating remediation of pollution, and demanding a moratorium on new industrial investment.

Wentworth, immediately adjacent to Merebank, was proclaimed a Coloured Group Area in 1963. "Coloured" people from various communities in Durban, and later from several provinces, were relocated here. The social ruptures of the removals created a community that was rootless and disorganized. Many men are fitters and construction workers who are employed intermittently. A relatively large proportion of households are headed by women, many of whom work in the service sector. In 1995, three previously rival civic groups linked with the ANC, AZAPO, and National Party formed the Wentworth Development Forum (WDF). Quickly, the WDF became active in negotiations with ENGEN and participated vigorously in many other environmental forums.

The Bluff, separated from these two Black communities by a buffer strip of greenspace, is a community of White civil servants, railroad workers, retirees, skilled blue-collar workers, and their families, of which about one-third are Afrikaans-speaking (an unusually high proportion in Durban). It is not as proximate to industry as Merebank and Wentworth, but the prevailing winds carry SO_2 fumes, oil sprays, and soot from the ENGEN refinery to homes on the Bluff. The Bluff does not have a tradition of community-based political mobilization, although the Bluff Ridge Conservancy and a chapter of the Wildlife Society have promoted conservation of the coastal biology and have addressed pollution issues, and several individuals confronted ENGEN intermittently in the 1970s and 1980s.

Since the early 1990s, organizations from these communities have built coalition structures that have become progressively wider and stronger. In April 1993, civic and environmental organizations from the Bluff and Merebank formed the South Durban Environmental Forum (SDEF) to address the air pollution of South Durban. Through participating in a profusion of environmental forums in 1994–1996, community organizations gained experience working together. Some of these forums also included organizations from nearby African and other Indian communities, significantly broadening the cross-community collaboration. Out of these experiences, an inclusive group of civic and environmental organizations formed the South Durban Community Environmental Alliance in late 1996, which has strengthened their collaboration and community outreach.

The ENGEN CAER Committee

In late 1994, consonant with the national discourse favoring consultation with affected communities, ENGEN invited selected individuals from these three nearby communities to join a Community Awareness and Emergency Response (CAER) Committee. The CAER Committee would acquaint the communities with the refinery's emergency procedures and create a forum to discuss their concerns. President Mandela had agreed to visit the refinery three months hence to honor the expansion of the refinery and its prospects for increased investment in Africa. ENGEN hoped that its apparent new openness, friendly meetings, and a promised opportunity to rub shoulders with the president would entice community members to quickly sign a CAER agreement that ENGEN could use to demonstrate that it had achieved consultation with its erstwhile critics.

The representatives of the CBEM saw President Mandela's impending visit as a unique point of leverage. A few weeks before the visit, the community representatives presented a detailed "Good Neighbour Agreement" to ENGEN. It included legally binding plans for reducing its emissions and its production of toxic chemicals. This was decidedly *not* the way the ENGEN management and its White consultants had intended the process to develop, and the company balked.

The Wentworth Development Forum decided to demonstrate at the gates of the refinery when Mandela came. Some of their placards read: "ENGEN is poison," "ENGEN is arrogant," "ENGEN stop killing us," "Our lives first, then profit," and "Comrade President, we request your intervention." As the president drove to the refinery gates, he left his limousine and spoke with the demonstrators. Then, he invited their representatives to meet with him the next day at his Durban residence.

Three days later, the president returned to Durban with three cabinet ministers for a longer meeting with community representatives, managers of the Durban refinery, and the ENGEN CEO from Cape Town, Rob Angel. At the meeting, Angel pledged that ENGEN had the will, the resources, and the intent to deal with the communities' problems.

General Bantu Holomisa, the then-ANC Deputy Minister of the Department of Environmental Affairs and Tourism (DEAT), was instructed by the president to convene a "South Durban Multi-Stakeholder Environmental Management Meeting" to deal with all the environmental issues there. This "Indaba" (a high-level consultation) raised expectations for changes among all the parties in South Durban and the city government and also took the spotlight off the ENGEN negotiations.

In intermittent discussions in the CAER Committee that followed, ENGEN presented its own draft agreement that called for a two-tiered process: first, agree to a Good Neighbourly Agreement that would provide a framework for consultation and for resolving problems (similar to its original CAER draft), and, later, develop action plans to address any environmental impacts of the refinery.

Several months later, ENGEN asserted that it would agree to all points except reducing SO_2 emissions, which clearly was the communities' key demand. ENGEN reluctantly released some documents to the CAER Committee and an engineer working with the CBEM reported that the documents showed that the refinery's emissions were approximately 10 times higher than would be allowable in similar refineries in the United States.

ENGEN delayed meetings for several months while its consultants researched SO_2 control policies of other industrialized countries. They returned to the negotiations insisting that there were no health problems resulting from SO_2 pollution and that the problems were largely a matter of "community perceptions." Repeatedly, ENGEN emphasized that its emissions levels did not violate South African law and, by implication, that the government DEAT agreed with the company that it was not necessary to reduce emissions in order to protect the health of local residents.

Community representatives wrote to President Mandela saying that, nine months after the meeting he had convened, no progress was being made with ENGEN on community grievances. Deputy Minister Holomisa convened another meeting, at which ENGEN called for an independent technical panel to assess the quantity and health effects of emissions of companies in South Durban. CAER Committee meetings stopped, and the technical panel never materialized. Some observers believed that bringing in technical experts from the two sides was unlikely to resolve the differences between the parties. Furthermore, in July 1996, Holomisa was replaced as DEAT Deputy Minister and expelled from the ANC, which created uncertainty about DEAT's role in the negotiations.

The South Durban Sulfur Dioxide Steering Committee

ENGEN had insisted repeatedly that it would reduce its SO_2 emissions only in accordance with a voluntary plan called the Bluff Valley Model (BVM), developed by the South Durban SO_2 Steering Committee, a regional committee of the major industrial emitters of SO_2 with national and local government representatives. Because ENGEN was using the BVM as its defense against investing in cleaner technology, some community representatives began to attend meetings of this committee. The BVM was a computerized system that used predictions of meteorological conditions and industry estimates of point-source emissions to predict when the ambient SO_2 concentrations at ground level were likely to exceed South African guidelines. Theoretically, the BVM was designed to enable companies to voluntarily reduce their SO_2 emissions during adverse weather conditions, especially from winter temperature inversions that could be expected to cause excessive ambient SO_2 concentrations.

Community representatives repeatedly argued that intermittent control of SO_2 through the BVM was not an acceptable solution to dangerous SO_2 emissions affecting their communities. Drawing on expertise from abroad, one of their arguments was that, 20 years ago, the U.S. EPA had rejected such intermittent control models as a means of meeting SO_2 emissions standards.

Despite ENGEN's repeated promises to comply with the BVM, during the winter of 1996 it did not switch to lower sulfur fuel when the BVM predicted poor air dispersion, arguing that there still was no management plan for the BVM, even though the SAPREF refinery voluntarily complied.

The SO_2 Steering Committee continued to operate the BVM model but could not agree on a management system. Various proposals were made for unblocking the process, including a lengthy conflict resolution process for which foreign funding was sought, but in which the community organizations refused to participate, saying that only reducing SO_2 pollution would resolve the conflict.

Participation in both the CAER Committee and the SO_2 Steering Committee gave the CBEM access to national political leaders and bureaucrats responsible for pollution control. Consequently, South Durban community representatives (along with civic groups from Cape Town engaged in negotiations with the Caltex refinery there) sought to influence national policy in meetings with national DEAT officials about revising oil refinery emission standards and in the CONNEPP process. Because of the total lack of progress on reducing emissions in two years of negotiations in both forums, the community organizations demanded that ENGEN develop a five-year plan to reduce its SO_2 emissions. If ENGEN did not respond within two months (by mid-1997), the CBEM threatened to withdraw from all forums in which ENGEN participated. To avoid this outcome, ENGEN submitted a document pledging to reduce its SO_2 emissions over time. While no specifics were given about either the amount of a reduction or the time horizon, ENGEN did back down from its insistence that it would act only on the basis of ambient SO_2 concentration and would not regulate its emissions.

THE DYNAMICS OF POWER IN ENVIRONMENTAL STRUGGLES IN SOUTH DURBAN

ENGEN remained recalcitrant for two years, even when faced with community negotiations and protests, significant negative publicity in the media, pressure from other industries and municipal government officials in the SO_2 Steering Committee, and exhortations from national ANC leaders.

However, the actions of the CBEM have not been a complete failure. They denied ENGEN a public-relations victory by refusing to sign an ineffectual CAER Committee agreement before President Mandela's visit, and they have kept the issue of ENGEN's pollution on the negotiating table and in the view of national policymakers. They demanded and obtained information from ENGEN, acquired experts and learned to interpret new technical data, organized demonstrations, took their story to the media, and threatened legal action. They built alliances among CBOs from all of the affected communities that were diverse in both race and class, which has contributed to their strength in other environmental forums. The CBEM also developed alliances with environmental activists in Cape Town, garnered the support of provincial and national environmental networks, and built relationships with oil refinery activists and environmental lawyers in the United States.

The Extralocal Sources of Power of ENGEN

ENGEN's refusal to acquiesce to the CBEM's demands is not surprising, given the imbalance of power between the company and the organizations in the surrounding communities. ENGEN's considerable political power in the local negotiations derived from both its national and multinational ties. Nationally, ENGEN is well connected with the DEAT bureaucrats retained from the old government who control air pollution. ENGEN successfully established a relationship with the new ANC leaders, as the visit of President Mandela demonstrated, and it reconstructed a new identity as the sole South African-owned oil refining company. It overtly supported the GNU and RDP, invested in green advertising, and publicized its motto: "We're South African—We Care."

In mid-1996, after two years of poor economic performance, ENGEN again became a multinational firm by selling a 30 percent controlling interest to Petronas, the Malaysian state-owned oil company. Ironically, this new alliance, born of economic weakness, has proven to be a source of political strength because Petronas' US$436 million investment in ENGEN was the largest foreign investment of the decade. Malaysian Prime Minister Dr. Mahathir Mohamed was a strong backer of the ANC during the liberation struggle and Malaysia was the source of half of all new foreign investment in South Africa in 1996, making it now the fourth largest foreign investor after Britain, the United States, and Germany (SAPA March 6, 1997).

It is difficult to distinguish the political significance of ENGEN's national and international linkages; indeed, they are not entirely separable. The ANC government is courting both national and foreign capital to expand investment and create the jobs that it assumes will develop with such investment. With both of these audiences, the government would not wish to engender negative assessment of either an inflexible labor market or tightened environmental regulations.

International Resources Appropriated by ENGEN and the Environmental Movement

The environmental forums we studied were explicitly designed as local consultations, but all of the actors used information, science, organizational linkages, political strategies, and alliances from beyond South Africa in order to enhance their resources and power.

The CAER Committee strategy that ENGEN employed had been developed in the early 1990s by North American chemical companies as one of several strategies for dealing with public opposition following the Bhopal accident and several U.S. chemical disasters. Several South African corporations affiliated with the Chemical and Allied Industries' Association of South Africa adopted this strategy. In 1993, because any communication by a corporation with civil society was perceived as progress, the chemical conglomerate AECI easily obtained community agreement to a CAER Committee without any demands for changing corporate practices. When, however, ENGEN formed its CAER Committee at the end of 1994, one of the community representatives had access to another model for community agreements with polluting corporations, the legally binding "Good Neighbor Agreement" developed by community organizations in the United States. The availability of this stronger model was a significant factor in the CBEM's adopting a tough negotiating stance.

Pollution standards were another topic on which both the refinery management and the CBEM looked abroad for legitimating norms and scientific knowledge. The CBEM used international standards, particularly from the United States, to argue its case that ENGEN's emissions were too high. ENGEN then contested those international standards and sought to demonstrate that South African standards were not out of step with some foreign norms (even though monitoring and sanctions for violations are virtually nonexistent in South Africa).

International contacts enabled the CBEM to contest the scientific and technical basis of corporations' defense of their technology, emissions, and standards for protecting community health.[10] Critiquing such arguments is generally difficult for locally based organizations because of the highly technical nature of the issues. It is even more difficult in semiperipheral countries such as South Africa, where there has been a culture of government and corporate secrecy and where there is a very small pool of people with expertise who are willing to serve the environmental movement rather than to mine the more lucrative fields of corporate and government consultancies.

The South African environmental movement also has received material assistance from foreign supporters. Foreign grants significantly increased the capacity of the MRA's environmental program in the early 1990s and more recently the new national Environmental Justice Networking Forum.

International linkages of the CBEM also were important as a source of solidarity and political support. Contact with U.S. activists who were fighting oil refinery pollution and environmental racism strengthened the resolve of the CBEM to maintain its demands for reducing pollution despite ENGEN's intransigence. In some campaigns, transnational links of communities and NGOs can focus public pressure on both the local management and the more distant parent company abroad. In this case, however, Malaysia—another country in the semiperiphery—does not have strong environmental standards, and the CBEM in South Durban has not been able to identify any organization in Malaysia that is concerned about Petronas' pollution.

Ambiguous Impacts of Democratization on the Power of the CBEM in South Durban

The new "culture of consultation" and the character of local democratization in South Africa have had ambiguous effects on the struggles over air pollution in South Durban. New consultative structures have enhanced the power of the CBEM but also have inhibited its ability to generate and use power, and the CBEM still has not achieved any reduction in SO_2 pollution.

On the positive side, the CBEM's participation in various new forums has established that South Durban residents must be regarded as stakeholders who have a right to be consulted on decisions about the environmental impacts of industry that affect them; however, the nature and extent of that role have been contested repeatedly.

Participation in forums sometimes has given the CBEM access to strategically important information and advance knowledge of decisions, which is a crucial resource for effective mobilization (Sandback 1980, 109). For example, ENGEN reluctantly divulged data about their emissions when the CBEM insisted that ENGEN do so as the price for the CBEM's continuing to participate in the forums. Old norms of secrecy and authoritarian rule change slowly, however, and Parliament has delayed action for two years on the Open Democracy Bill that would provide wide ranging access to government information. In these South Durban cases, obtaining both corporate and governmental information generally required repeated demands.[11]

On the negative side, on several occasions, organizations in the CBEM questioned whether continuing to participate in forums any longer furthered their agendas and merited the large investment of time and energy that they required. As Gould et al. (1996, 36) found in their U.S. case studies, transnational corporations "can frequently wait out and wear down their opponents. These battles are clearly wars of attrition."

Some national and municipal officials and consultants have pressed the CBEM to remain in consultative forums and moderate their demands if necessary in the interest of maintaining democratic participation *per se,* insisting that the CBEM must give up the "old culture of confrontation" and embrace the "new culture of consultation." Their position implicitly ignores the imbalance of political power between corporations and the CBEM and the different ways in which they generate power. Offe posits that unions (and, we would argue, social movements as well) at certain times must operate in more visible ways and must take the offensive. Their power depends upon the collective action of their members, such as public protests, or at least the potential for such action.

The CBEM in South Durban has refused to abandon its right to organize public protests as a price for remaining in consultations. Strategic use of public protests was crucial in the ENGEN campaign and several others.[12] Organizations in the CBEM also refused to agree to corporate and government demands that they not speak independently to the media, which is a crucial vehicle for mobilizing their members and potential allies and for delegitimating corporate claims and propaganda.

The CBEM sometimes has had the power to deny legitimacy to a consultative process by refusing to participate in it, provided that it shepherds this power and uses it sparingly. The CBEM has threatened to act or actually used this power several times in South Durban, including in the SO_2 Steering Committee.[13] However, the honeymoon period of local government's commitment to forceful community participation was short-lived, as several officials recently criticized community leaders for "being too pushy," including their requesting information on city budgets for environmental consultants. Now they also insist that various environmental assessments and industrial planning can go ahead "on the fast-track" even if the CBEM refuses to participate.

CONCLUSION

Increased integration into the global economy and democratization have generated immense effects on South Africa's economy and its polity. They have reshaped the parameters within which local communities are addressing the accumulated environmental problems caused by unbridled state-protected strategic industries during apartheid. This has led to a turbulent period when, encouraged by the new democracy, communities are voicing high expectations for change and corporations are testing the water to see how much change they must tolerate and social overhead capital they must expend in order to achieve political legitimacy in the new South Africa. Simultaneously, the managers lobby government for the new neoliberal dispensation and against significant new pollution control.

The community-based civic and environmental organizations in South Durban that we studied are part of a growing national progressive environmental movement in South Africa. The CBEM in Durban and others in Cape Town, Port Elizabeth, and the Vaal Triangle that are waging pollution campaigns against industries have become well aware that the environmental problems they face are caused by firms based outside their cities or even outside South Africa.

Another stream in this national movement has focused on mobilizing nationally against toxic and nuclear wastes. Earthlife Africa has been closely linked to peace and justice, antinuclear, and environmental movements in the core nations. It has mounted campaigns in South Africa (including in Durban) against international transportation of nuclear waste around South Africa's coast. Earthlife also has lobbied against importing toxic waste into South Africa from neighboring countries and from core capitalist countries that have more stringent environmental regulations.

Nationally, the progressive environmental movement has coalesced since 1994 in the Environmental Justice Networking Forum (EJNF). EJNF has become a broad, multiracial network that assists its affiliates to respond to the many different environmental issues facing both African rural and urban communities, as well as to larger questions of national economic and environmental policy. EJNF has strong working relationships with the labor and civic organizations that are key components of the broader South African progressive social movement. EJNF has become a major voice for democratic participation in strengthening South African environmental policy and in opposing foreign-driven economic policies of the World Bank in South Africa and other African countries.

Some of the "green," recycling, and antilitter organizations in South Africa have joined the progressive environmental social movement. A few of these organizations have lobbied for stronger national environmental policy and have participated in local antipollution campaigns, occasionally as a force of moderation and other times following the lead of the CBEM. However, many "green" organizations are heavily funded by major South African industrial (especially petrochemical) and mining companies that publicize such grants to "green" their own image.

While the sources of investment, financing, management, technology, and trade are largely transnational, many of the environmental struggles are being waged at the local level. In this grossly unequal balance of power, CBEMs have sought to increase their power by utilizing both the mandate for consultation that is at the core of the political discourse of the new government as well as the strong tradition of independent action that the South African civic movement built during the antiapartheid struggle. They also have strengthened their alliances across racially and economically diverse communities, in national environmental and political arenas, and with movements abroad.

The transnational character of the polluters has not worked entirely against the environmentalists, however. For example, Thor Chemicals Holdings, a British multinational, moved its processing of mercury products from a plant in southeast England to KwaZulu-Natal Province in South Africa, where several workers subsequently died of mercury poisoning. When South African courts granted only a paltry settlement, EJNF worked with a lawyer in the United Kingdom who won a sizable award when the British court ruled that it had jurisdiction over Thor's negligence in its South African operation. Working at both the national and transnational levels to develop creative political and legal strategies to the environmental problems of these South African communities holds great promise.

The state also is an actor in these environmental struggles. The CBEMs have learned from experience that government cannot be counted on to be an ally, or even to be neutral, when local communities confront powerful corporations. The national and local structures of the state are seeking a *via media* that threads its way between the expectations and demands of long-contending parties. The coming to power of the ANC-led government unquestionably creates more political space and political legitimacy for social movements, including the environmental justice movement. However, the state also seeks to win investor confidence in an increasingly competitive world in order to ensure the growth of production and jobs. A robust economy is needed to fuel the state's treasury for addressing the panoply of pent-up basic human needs. Pressures from Western governments, international financial institutions, and the global market legitimate foreign-driven, capital-intensive strategies for development and curtail impulses to regulate industry for environmental and human protection or to transfer the costs of the externalities of industrial pollution to distant owners.

Organizations in the environmental movement that utilize the structures of consultation frequently have confronted the limits of democratic participation. The CBEMs are woefully under-resourced for these struggles, often lacking funds for taxi fare to meetings, lost wages, or even English-to-Zulu translations. Government agencies frequently have been more willing to fund yet another consultant's report to study the perceptions of people in affected communities (without conferring with those communities directly about the usefulness of the expenditure) than to provide funds to the communities to organize on their own needs.

Despite a rhetoric favoring "sustainable development," economic growth and environmental protection have operated largely on two separate tracks, and the environment has taken a back seat to economic growth. Despite Durban's formal commitment to Local Agenda 21 (LA 21), environmental activists there report that they are excluded from key conversations in planning industrial expansion and commercial development and that their concern about pollution reduction and cleaner technology is regarded as obstructionist. As CBOs and NGOs fill the new political space that was created by the ANC government, some politicians and bureaucrats try to curtail that space. Some officials are impatient with the delays, questions, interventions, and processes that occur in local participation; as a result, they frequently do not want their authority questioned.[14]

Environmentalists express concern about lack of local participation and thorough environmental assessment in the fast-track SDIs. They include the SDI proposed for Durban in May 1997 which would expand South Durban's liquid fuels and chemical industry cluster, leading some environmentalists to speculate that South Durban is becoming a "sacrifice zone—the Baton Rouge of South Africa." Economic factors more than any other consideration circumscribe government policy on the environment. As in many other semiperipheral states, the impetus for economic growth and the competition for transnational investment legitimizes postponing expenditures for effective strategies for environmental regulation and sustainable development.

Some of the struggles against major companies, such as those in South Durban, have demonstrated to the CBEM that, for local democracy to work, environmental activists must generate influence sufficient to affect national policy and political leaders. As Gould et al. (1996, 37) found in their study of remediating pollution in the North American Great Lakes, "effective environmental resistance to the treadmill [of expanding capitalist corporations] requires strong local opposition, as well as support from political centers of power."

While ownership and production are increasingly transnational, new lines of communication are available to link environmental movements transnationally across the semiperipheral and core countries. This allows them to mobilize together to develop sophisticated strategies to oppose those corporate practices that spread the environmental costs of production across the globe and to influence the state to accede to their demands for environmental justice.

NOTES

Our research began as part of a project by the Institute for Social and Economic Research of the University of Durban-Westville, under contract by the City of Durban Environmental Manager. This contract provided support for a portion of the research conducted by Sven Peek. David Wiley received financial support from Michigan State University, an MSU All-University Research Initiation Grant, and the Fulbright-Hays Faculty Research Awards Program of the U.S. Department of Education. Seyathie Ramurath also was a member of this research team in 1995–1996 and provided important insights about environmental issues and movements in South Durban. We offer our especial thanks to Professors William G. Martin and James Mittelman for their reading and valuable comments on earlier drafts of this chapter.

1. The other cases we studied involved negotiations leading to the successful closing of a hazardous waste site in an African community, establishing an environmental-management committee for South Durban, conducting an environmental-impact assessment for a new plastics factory, efforts to clean up industrial water pollution and rehabilitate the Isipingo estuary, and clean up after the accidental poisoning of a group of toddlers at a closed chemi-

cal plant in Wentworth. These cases are discussed in Wiley et al. (1996), Wiley and Root (1996), and Root and Wiley (1996).

2. The organizations that were included in the South Durban environmental movement include: Athlone Resident's Committee, Bluff Ratepayers Association, Bluff Ridge Conservancy, Isipingo Environmental Committee, Lamontville SANCO, Merebank Enviro-Watch, Merebank Residents Association, Umlazi Simunye Youth Development Club, Wentworth Development Forum, Earthlife Africa (Durban), and The Wildlife Society (Durban and the Bluff). These organizations formed the South Durban Community Environmental Alliance in December 1996.

3. Article 24 of the Constitution reads: "Everyone has the right—(a) to an environment that is not harmful to their health or well-being; and (b) to have the environment protected, for the benefit of present and future generations, through reasonable legislative and other measures that—(i) prevent pollution and ecological degradation; (ii) promote conservation; and (iii) secure ecologically sustainable development and use of natural resources while promoting justifiable economic and social development."

4. See, for example, Gould's study of government regulators' sympathy toward a paper mill emitting toxins into Lake Michigan and Lindblom's study of power exerted by U.S. Steel over environmental regulations in Gary, Indiana because these firms dominated those local economies (Gould et al. 1996; Garner 1996). See also Cahn's critique of U.S. environmental regulations that are primarily symbolic and are largely not implemented (Cahn 1995).

5. The Manila Declaration on People's Participation and Sustainable Development, authored by 31 NGO leaders in June 1989, is one important example (Korten 1990, 217).

6. From the perspective of the ANC, such a commitment to participatory democracy was necessary in order to maintain political support from the organizations of civic society that had played a crucial role in waging the antiapartheid struggle and bringing the ANC to power.

7. Racial groups are not fixed social categories but are labels and symbolic categories. Nevertheless, the racial typology of the old South Africa endures in the social structure of the country; therefore, we utilize terms for the "racial" categories designated by the apartheid government and use the term "Black" to refer to all racial groups other than Whites.

8. The Legal Resource Centre in Cape Town has noted that "over the past 10 to 15 years the knowledge of the health and environmental impacts of these pollutants has increased dramatically" and that the standards governing their emissions need to be tightened considerably (Legal Resources Centre 1996).

9. The National Key Points Act has not yet been repealed, and it continues to have disastrous consequences for local communities in denying them access to key environmental information. For instance, on 29 April 1997, 160,000 liters of crude oil spilled from a ruptured pipeline into a local river near Newcastle. Because it was covered by the legislation, local residents were not informed until eight days later.

10. In another case of an accidental poisoning on the site of the Chemico Ltd. plant (a subsidiary of ENGEN) immediately adjacent to low-income flats in Wentworth (Austerville), the Wentworth Development Forum was able to access a U.S. university toxicology information service via the internet to force ENGEN, against its wishes, to totally eliminate toxic lindane from a toddlers' sandbox in the playground next to the abandoned factory instead of leaving substantial residues there. Lindane is a highly toxic pesticide now banned in the United States.

11. In another one of the South Durban forums, information gained from the consultation process was very important. The CBEM learned about the Waste-tech company's plans to expand its hazardous waste site in nearby Umlazi because of new national regulations

requiring local consultation. This knowledge enabled the communities to expand and more effectively target their organizing efforts against the site.

12. Public protests by several hundred Indian and African students—supported by their parents, teachers, and school administrators—played a decisive role in the campaign to close the hazardous waste site in nearby Umlazi.

13. In another South Durban forum, on several occasions CBOs threatened to leave the monitoring committee on the Waste-tech hazardous waste site in order to call the attention of national regulatory authorities to the flawed consultation taking place there.

14. For example, at the CONNEPP II conference in January 1997, the head of the Environment and Agriculture Department in Gauteng province stated baldly: "I did smash the Provincial Environmental Advisory Forum immediately when I came into power, and that was the correct choice" (CONNEPP 1997).

REFERENCES

Cahn, Matthew Alan. 1995. *Environmental Deceptions: The Tension between Liberalism and Environmental Policymaking in the United States*. Albany: State University of New York Press.

Chetty, Siva, and Sven Peek. 1995. Presentation by Siva Chetty of the Merebank Residents Association and Sven Peek of the Wentworth Development Forum. In *Group Reports and Proceedings of the South Durban Multi-Stakeholder Environmental Management Meeting held in the Durban City Jubilee Hall on May 4, 1995*. Lombard and Associates.

Consultative National Environmental Policy Process (CONNEPP). 1997. Proceedings of CONNEPP II Conference, 24–25 January 1997.

Diab, Roseanne, and Rob Preston-Whyte. 1996. Air. In *Report on the State of the Environment and Development of the Durban Metropolitan Area,* Volume 2, compiled by Doug Hindson. Submitted to Executive Director (Physical Environment Service unity), North Central Council, Durban Metropolitan Area.

Edmunds, Marion, and Rehana Rossouw. 1997. Gear under Threat. *Weekly Mail and Guardian,* January 17.

Garner, Robert. 1996. *Environmental Politics*. New York: Prentice-Hall/ Harvester Wheatsheaf.

Gould, Kenneth A., Allan Schnaiberg, and Adam Weinberg. 1996. *Local Environmental Struggles: Citizen Activism in the Treadmill of Production.* Cambridge, UK: Cambridge University Press.

International Council for Local Environmental Initiatives (ICLEI) and the Durban Local Agenda 21 Working Team. 1995. Case Studies on the Local Agenda 21 Process: Durban. Revised October 1995 (http://www.iclei.org/csdcases).

Kistnasamy, M. B. 1994. The Relationship between Location of Residency and Respiratory Symptoms of Primary School Pupils. Unpublished manuscript.

Korten, David C. 1990. *Getting to the 21st Century: Voluntary Action and the Global Agenda.* West Hartford, CT: Kumarian Press.

Legal Resources Centre, 1996. Proposal for Reductions in Atmospheric Pollution for the Oil Refining Industry. Unpublished manuscript. June 20.

LeRoux, Pieter. 1996. The State and Social Transformation. In *Reconstruction, Development and People,* ed. Jan Coetzee and Joann Graaff. Johannesburg: International Thomson Publishing.

Mandela, Nelson. 1995. Opening Address by President Nelson Mandela at the Conference on National Environmental Policy, Johannesburg, 17 August 1995. Issued by: Office of the President (gopher: //gopher.anc.org.za:70/00/ govdocs/speeches/1995/sp0817.01).

Mandela, Nelson. 1996. Address by President Nelson Mandela on the Occasion of the 75th Anniversary of the South African Communist Party, Cape Town 28 July 1996. Issued by: African National Congress (gopher: //gopher.anc.org.za:70/00/govdocs/speeches/1996/sp0728.01).

Mayekiso, Mzwanele. 1996. *Township Politics: Civic Struggles for a New South Africa.* New York: Monthly Review Press.

Modavi, Neghin. 1996. Mediation of Environmental Conflicts in Hawaii: Win-Win or Cooptation? *Sociological Perspectives* 38(2): 309–16.

Offe, Claus. 1985. *Disorganized Capitalism: Contemporary Transformation of Work and Politics.* Cambridge, MA: MIT Press.

Padayachee, Vishnu. 1996. The International Monetary Fund and World Bank in Post-Apartheid South Africa: Prospects and Dangers. In *Reconstruction, Development and People,* ed. Jan Coetzee and Joann Graaff. Johannesburg: International Thomson Publishing.

Petrie, J. G., Y. M. Burns, and W. Bray. 1992. Air Pollution. In *Environmental Management in South Africa,* ed. R. F. Fuggle and M. A. Rabie. Cape Town: Juta & Co.

Ramaphosa, Cyril. 1996. Address at the South Africa Chamber of Business Annual Banquet, 28 May. Issued by: African National Congress (gopher: //gopher.anc.org.za:70/00/anc/speeches/1996/sp0528.01).

Root, Christine, and David Wiley. 1996. Community Participation and National Regulation in the New South Africa for Managing Hazardous Industrial Waste. Presented at the 1996 Annual Meeting of the African Studies Association, San Francisco.

Sandback, Francis. 1980. *Environment Ideology and Policy.* Montclair, NJ: Allanheld, Osmun Publishers.

Scott, Dianne. 1994. Communal Space Construction: The Rise and Fall of Clairwood and District. PhD diss., Department of Geographical and Environmental Sciences, University of Natal, Durban.

South African Press Agency (SAPA). 1997. Mandela Arrives to Bolster Ties with Malaysia. March 6.

Southern African Economist. 1988. The Great Trek of the Multinationals. 1(1): 14–15.

United Nations Conference on the Environment and Development (UNCED). 1992. *Agenda 21.* Chapter 28: Local Authorities Initiates in Support of Agenda 21 Programme Area.

Whyte, Anne, ed. 1995. *Environment, Reconstruction, and Development.* Volume 4 of *Building a New South Africa.* Ottawa: International Development Research Center.

Wiley, David, and Christine Root. 1996. Environmentalism in the New South Africa: Conflict and Cooperation in Durban's Petrochemical Basin. Paper presented at the 1996 Annual Meeting of the African Studies Association, San Francisco.

Wiley, David, Christine Root, Sven Peek, and Seyathie Ramurath. 1996. Negotiating Environment and Development in South Durban: Communities, Industries, and Authorities. A Report to the Environmental Manager of the City of Durban, July.

12

The Emergence of South Korean Environmental Movements: A Response (and Challenge?) to Semiperipheral Industrialization

Su-Hoon Lee & David A. Smith

> Dark smoke arising from factories is a symbol of our nation's growth and prosperity. (President Park Chung Hee 1962)

> Pollution is a phenomena that appears when the contradictions of capitalism, with its sole objective of pursuing profits, reach their extreme. (Korean Pollution Research Center 1982)

> A nation-state will quickly lose its legitimacy if citizens can no longer breathe the air or drink the water. (Kenneth Gould, Allan Schnaiberg, and Adam Weinberg 1996, 17)

In the last three decades South Korea has undergone one of the most rapid industrial transformations in world history. Since the early 1960s the economic growth has been remarkable: overall GDP grew at 8.6 percent between 1960 and 1970, and 9.5 percent over the following decade, while sectoral growth in industry and manufacturing exceeded 15 percent for that entire period (data from World Bank 1982, summarized by Hart-Landsberg 1993). Carter Eckert (1992, 289) provides an evocative description of this process as

> a tale whose drama is heightened by breathtaking contrasts: a per capita GNP of about US$100 in 1963 versus a figure of nearly US$5,000 as the year 1990 began; a war-ravaged Seoul of gutted buildings, rubble, beggars, and orphans in 1953 versus the proud, bustling city of the 1988 Summer Olympics with its skyscrapers, subways, plush restaurants, boutiques, first-class hotels, and prosperous middle class; a country abjectly dependent on foreign aid in the 1950s versus a 1980s economic powerhouse. (Quoted by So and Chui 1995, 191)

It is hardly surprising, then, that many Western scholars, particularly those in the United States, describe South Korean development as an "economic miracle" and suggest that the Korean model be adopted by other, less-developed Third World societies (for an influential example, see Balassa and Williamson 1987).

Accepting this premise, scholars and policymakers have honed in on the South Korean case, along with the other so-called East Asian Newly Industrialized

Countries or Economies (NICs or NIEs), attempting to discover the secret of this "success." The literature in this area has grown almost as fast as the South Korean economy: for instance, a vibrant discussion developed in the late 1980s contrasting the East Asian NICs to Latin America (see Evans 1987; Gereffi and Wyman 1990), while others explained the critical role that East Asian states played in the rapid economic growth (Amsden 1989; Wade 1990; Appelbaum and Henderson 1992; Evans 1995). By debunking simplistic neoclassical images of the East Asian "economic miracles" as triumphs of free-market capitalism and detailing the complex ways that political and economic networks and institutions promoted rapid growth, this work significantly improved our analytic grasp of the transformation of South Korea and the other nations of the Asian Pacific Rim.

Nevertheless, even these analyses share a characteristic of the vast majority of scholarly and popular images of South Korea, portraying the economic transformation as a great success, to be explained, and, perhaps, to be emulated. Interestingly, many Korean social scientists take a decidedly more critical view of the South Korean economic development, highlighting the constrained nature of "dependent development" (Lim 1985) and the problematic nature of growing regional inequality (Chon 1992) or the power of the giant *chaebol* (Kim 1997). Concurring with this more skeptical view are some Western scholars who are most intimately familiar with the South Korean case, like Bruce Cumings (1984, 1989) and Martin Hart-Landsberg (1988, 1993). Our own previous research led us to conclude that, in terms of both political democratization and continued economic dynamism, the South Korean "semiperipheral success story" faced potential limits and constraints (Smith and Lee 1990, 1991a).

Clearly, while general assessments of the South Korean development trajectory vary widely (and are indelibly colored by theoretical assumptions or worldviews of the commentators), it is difficult for anyone to deny that the high velocity of industrialization has had some negative consequences. Among these is severe and irrevocable damage to the natural environment. This situation, which could fairly be called an ecological crisis, is the backdrop for our discussion.

Often people refuse to recognize this dirty underside of rapid industrialization. In fact, for many years Korean-government technocrats assumed, perhaps with the tacit consent of the population, that environmental damage is an unavoidable and necessary by-product of fast-track economic growth. President Park's words equating "dark smoke" with "growth and prosperity" vividly convey the ecological insensitivity that dominated official thought and discourse. Nevertheless, by the 1980s, the South Korean populace was beginning to awaken to the severity of the ecological crisis that they faced, as they became increasingly aware of the devastating toll that industrialization and rapid growth had exacted on the natural environment. Like increasingly ecologically conscious people in other parts of the world, the Korean citizenry began to question the growth-oriented, achievement-oriented, target-oriented model of development championed by politicians, bureaucrats, and businessmen and orchestrated by the state. This questioning led to attempts to take concrete actions to prevent further despoilage of the environment and reverse burgeoning levels of pollution. Over time this ecologically based

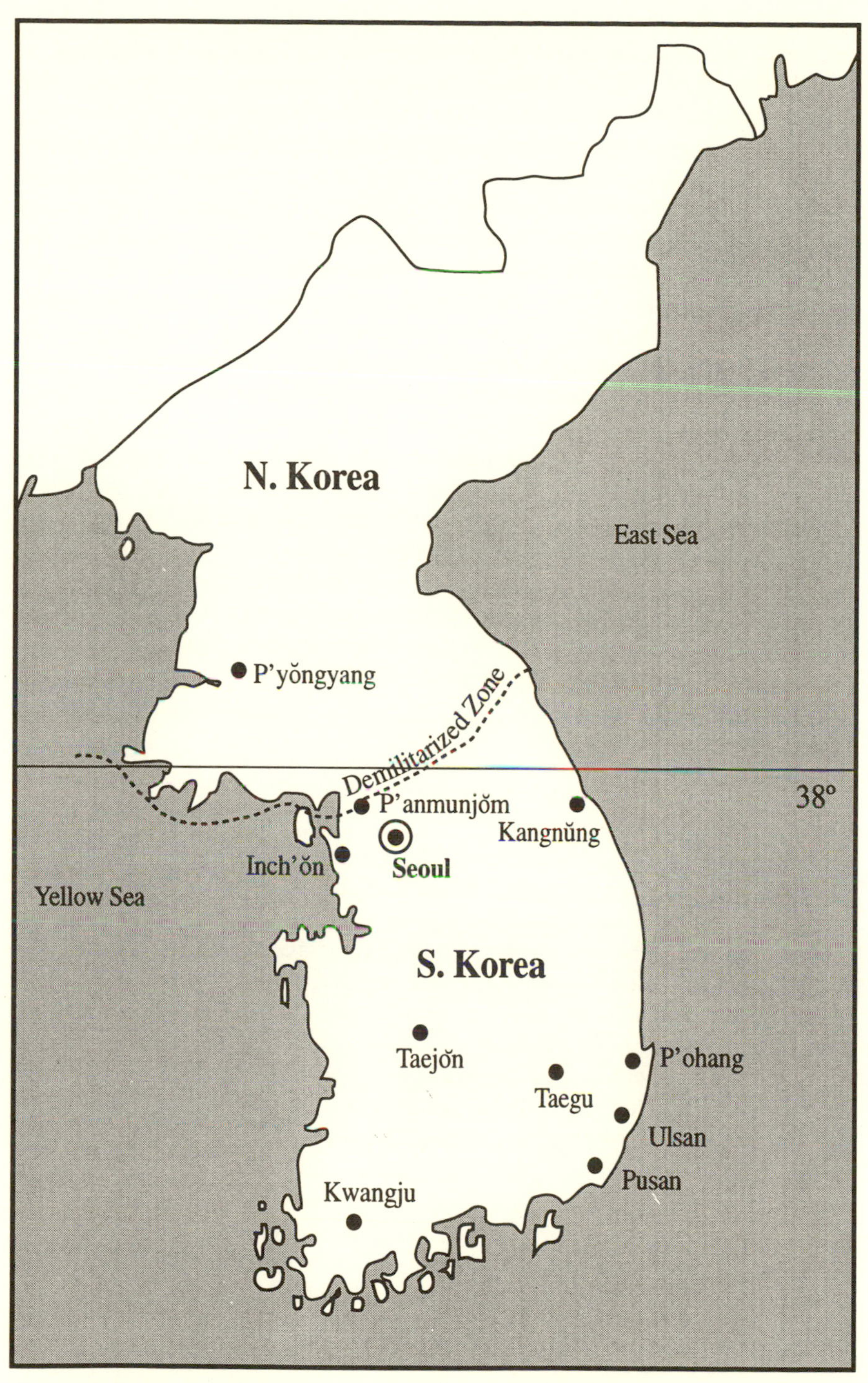

Map 12.1 North and South Korea

opposition to "business as usual" became increasingly organized, signaling the emergence and consolidation of the environmental movement in South Korea.

In this chapter we will examine the recent history of various strands of this nascent organization of environmentalism as a social movement and try to understand it as a response (and, perhaps, a challenge) to semiperipheral industrialization. The following section attempts to place South Korean environmental organization into an appropriate theoretical framework, drawing on concepts from world-system analysis, notions about "new social movements" and environmental sociology, and synthesizing Western and Korean literatures. Next, we sketch the extent of ecological destruction wrought by the Korean "economic miracle." The bulk of the chapter consists of a short history of environmental activism in South Korea in the past two decades, providing an overview of the various local and national strands of organization, as well as some indication of the manner and degree to which the "movement" has coalesced into a coherent actor on the national stage. What are the key issues, major activities, and notable achievements? What direct and/or indirect linkages does Korean environmentalism have to global ecological groups, ideologies, and networks, and international watersheds like the Rio Conference? What linkages do Korean "green" organizations have to other social movements pursuing goals of democratization or economic justice? Answering these questions should provide a sense of how the South Korean environmental movement is embedded in global political economy, its connections to worldwide ecological struggles, and the degree to which can be described as "antisystemic."

THEORETICAL CONTEXT

As a semiperipheral NIC experiencing very rapid industrial growth, South Korea is a strategic case for research on environmentalism and other potentially antisystemic "new social movements." The country has undergone a distinctive pattern of "dependent development" based on the rapid growth of export manufacturing and contingent on a particular constellation of relationships between the state, capital, and labor over several decades. South Korea's particular niche in the world-system and "global commodity chains" (see Gereffi and Korzeniewicz 1994) conditions the nation's class structure, politics, and future development options. Like other semiperipheral states, it is prone to distinctive, particularly volatile, social and political conflicts and contradictions (Wallerstein 1979; Mouzelis 1986).

Indeed, of all the East Asian NICs, South Korea has the most polarized class structure and most confrontational politics, with a heritage of radical protest by students and workers (Deyo 1989). The mid- to late-1980s witnessed a surge of popular mobilization (union organization, strikes, mass demonstrations) that suggested that these movements had some radical potential to threaten state and corporate power (Smith and Lee 1990, 1991a). Yet in the 1990s the pendulum seems to have swung back toward mass quiescence, demobilization, and top–down political control.

Not surprisingly, our first attempt to understand the emergence of an environmental movement in South Korea speculated on the degree to which these organizations might be "antisystemic" and build coalitions with other oppositional movements (Smith and Lee 1991b). Choi (1993) argues that Korean environmental-movement groups *are* antisystemic, at least to the extent that they are the creations of, and a reaction to, Korean capitalist development. But both Choi (1992) and Kim (1991) also highlight salient differences between groups focussed on environmental concerns. Kim (1991) sees a clear distinction between "progressive" environmental organizations and "conservative" ones, in terms of both their conceptions about the nature of ecological problems and their tactics. Similarly, Choi (1992) looked at whether movements were primarily "social" or "ecological," "local" or "nonlocal." Based on this classification he found four categories of movements organized around different foci: 1) local resident, 2) antipollution, 3) livelihood/community, and 4) environmental protection.

Both the relatively heterogenous class-basis of South Korean environmental activism and the green movement's difficulty linking up with other oppositional movements suggest that it may be useful to view it as a possible "new social movement" (Eder 1993). S. J. Lee (1992) and Ku (1994) both argue that Korean environmentalism falls into this category. They claim that these groups and their membership have emergent ecological concerns, distinct from economic or class interests, and are using new stratagems. To Ku (1994), whose work represents the most extensive documentation of the environmental movement thus far, this suggests affinities to Western "new social movements." This concept, developed to explain recent changes in advanced (post)industrial societies, identifies new oppositional movements that are very different from old class-based organizing, in that they are largely supported by "the new middle classes" of young, educated, service and professional workers (Offe 1985; Eder 1993). These groups are in positions where they are more likely to focus on "lifestyle" issues that are not linked to material economic issues. Influential analyses of European environmentalism focus on the extent to which they constitute "new social movements" reflecting "postmaterial values" (Kitschelt and Hellemans 1990; Dalton 1994; Diani 1995). This approach stresses the key role played by growing middle and professional classes in supporting contemporary environmentalism, leading to an emphasis on increasingly "moderate" mainstream goals—an empirical pattern we find in the South Korean case. Conceiving of environmentalism in terms of "new social movements" (or for that matter, using the discourse of "ecofeminism" a la Mies (1986) or Shiva (1988) may provide special insights into the normative and ideological motivations of participants and provide a cautionary lesson about the degree to which ecological concerns link with class-based politics.[1] On the other hand, the rather amorphous and eclectic nature of this approach may limit its usefulness as a theoretical tool for understanding the political and economic implications of (and the cleavages within) environmental movements in Korea and elsewhere.

One issue that *must* be dealt with in the South Korean case is the tension between the moderation and "insider politics" focus of Seoul-based, national environmental organizations and the dynamism of local community actions across the southern half of the peninsula. Fortunately, an analogous bifurcation has already been extensively studied by students of contemporary U.S. environmentalism.

Gottlieb (1993) clearly contrasts the "professionalization and institutionalization" of mainstream American environmental groups, with the "direct action" of alternative "grassroots" movements. He describes divergent trajectories in the 1980s, as the mainstream "group of Ten" organizations became "players" on the Washington political stage, but blunted their adversarial edges and risked cooptation, while alternative citizens' movements coalesced and mobilized around local and community ecological issues. Szasz (1994) focuses on one of these alternative citizens' campaigns: the antitoxic waste movement in the United States. Emphasizing their dynamism (and potential to generate "a politics that works"), Szasz recounts how local activists and groups, ostensibly starting from a parochial and phobic "Not in My Backyard" (NIMBY) standpoint, often moved toward an encompassing vision of environmental and social justice, which he characterizes as "radical environmental populism." This movement emerging from grassroots environmental organizing, Szasz argues, could help revitalize progressive politics in the United States, as these groups develop links to labor, feminist, and civil rights organizations; similarly, Gottlieb (1993) suggests that "environmentalism redefined," propelled by alternative grassroots movements, may provide hope for broad-based political activism and change.

Environmental sociology grounded in the political-economy tradition offers a final (if somewhat less sanguine) literature to mine for useful ideas. The approach to environmental struggles developed by Allan Schnaiberg and his associates offers one useful departure (Schnaiberg 1980; Schnaiberg and Gould 1994). Modern industrial society is composed of self-perpetuating, intensifying "treadmills of production" whose inexorable accelerations (motivated by the quest for expanding profits) lead to progressively greater ecological damage and depletion. In the most recent explication, they emphasize the global dynamic in a theoretical chapter entitled "Transnational Structures and the Limits of Resistance" (Gould, Schnaiberg, and Weinberg 1996, chap. 1). The citizen-workers and community activists they study are emboldened to oppose the corporate "treadmill" when "their *use-value* interests in local ecosystems" (15, their emphasis) are threatened by land development and pollution.[2] But citizen-worker struggles to protect their local environments, which were always uphill battles, become even more difficult in the contemporary "transnationalized" economy. That's because in the current "post-Fordist" era of global neoliberalism in which manufacturing firms can relocate to peripheral countries, there is an increased push toward "low road" cost-saving strategies that lead to pressure to reduce wages and lessen environmental regulation and enforcement. Gould, Schnaiberg, and Weinberg (1996) argue that states are caught in a contradictory dynamic between facilitating growth and maintaining a sustainable environment (at minimum air to breathe and water to drink and, hopefully, a good deal more). However, they also recognize that the battle to preserve ecological use-value will go on at the local and regional level.

This emphasis on environmental action that occurs in local communities motivated by residents' interests to protect and conserve nature "where they live" (consistent with Gottlieb [1993] and Szasz' [1994] descriptions of U.S. grassroots environmentalism) suggests another grounded political-economic approach. Both

U.S. and Korean environmental sociologists point to Karl Polanyi's insights into the "double movement" that characterized the rise of market societies, as a tool for understanding the impetus for environmental movements (Foster 1994; Han 1994). Polanyi (1957, 132) describes

> two organizing principles in society. . . . The one was the principle of *economic liberalism*, aiming at the establishment of a self-regulating market relying on the support of the trading classes and using largely *laissez-faire* and free trade as its methods; the other was the principle of *social protection* aiming at the conservation of man and nature as well as productive organization, relying on the varying support of those most immediately affected by the deleterious action of the market—primarily, but not exclusively the working and landed classes—and using protective legislation, restrictive associations, and other instruments of intervention in its methods. (Quoted by Foster 1994, 134–35)

Unlike the "new social movements" conceptualization, understanding local people involved in environmental movements as struggling to promote Polanyian "social protection" (or to conserve the use-value of their local natural environment) keeps our focus on material conditions and class alignments. At the same time, though, these environmental–sociology perspectives explain why ecological concerns tend to crystallize at the local or community level in South Korea and other societies. But they also retain an acute awareness of the need to contextualize environmental problems as "transnational" or in world-system dynamics. However, figuring out ways to link "local" to "national," much less "global," is much more difficult than facile sloganeering implies.

SOUTH KOREA'S ENVIRONMENTAL CRISIS

Ecological awareness and environmental movements developed in response to objective conditions. The high speed of industrialization in South Korea over the past three or four decades has exacted a terrible price on the natural environment. The ecological damage is extensive and well-documented. In the last decade the environmental crisis became widely understood by the Korean people and openly acknowledged by elites in business and government. In fact, by 1986 the Korean Pollution Research Institute listed three defining features of environmental degradation in the country: first, it is nationwide and affects every individual and community to some degree; second, pollution is increasing very rapidly, as more toxins are produced as industrial by-products; and third, pollution is of a highly malignant nature that fundamentally threatens people and ecosystems, with the deadly nature of some of these noxious chemicals only gradually becoming obvious. In the following section we provide a quick overview of the nature and extent of pollution problems that affect the entire population.

Air Pollution

Air quality is the most obvious element in the urban environment that has deteriorated dramatically during the period of rapid industrialization. There is atmospheric pollution in every city (large or small) in South Korea, with air pollution reaching extremely unhealthy, almost intolerable levels in Seoul and Pusan, where smog is an obvious reality to any resident or visitor (Eder 1996, 68). Among the world's major cities in the late 1980s, Seoul had the dubious honor of scoring the highest on sulphur dioxide in the atmosphere: in 1989 the average annual sulphur dioxide level reached 0.061 parts per million (ppm). The inflated tolerance level set by the South Korean government is 0.05 ppm (compare that to the U.S. EPA standard of 0.03 ppm). Eighty-three percent of annual rainfall in Seoul contained such high acid levels (PH less than 5.6) that it was considered a threat to human health. Dust pollution is also a major problem. In 1988, the density of particulate pollution was 179 milligrams per cubic meter (mpcm), which also exceeds the government's tolerance level of 175 mpcm (U.S. EPA standards currently are 50 mpcm and critics maintain that level is too high!). Among the major causes of declining air quality in Seoul are motor vehicle emissions, heating with charcoal, and the burning of Bunker-C oil. By the end of 1989, the number of automobiles in Seoul was over one million, and traffic volume has increased *dramatically* since then. Families in poorer neighborhoods continue to rely on charcoal fires for cooking and winter heating. While authorities have made some effort to improve air quality recently, smog remains a major problem.

Water Pollution

Plainly put, South Korean municipal water is not safe to drink. In August 1989 government investigators discovered that water at ten purification plants in the city contained heavy metals, such as iron, manganese, and cadmium, at up to twice the official tolerance levels (Bello and Rosenfeld 1990, 101). The next summer a major newspaper reported that trihalometan (THM) was present in Seoul's municipal water. In a survey taken about this time, 98 percent of Seoul residents reported that their household tap water was not drinkable (*Hangyeore Shinmun,* July 10, 1990). In the following months, newspapers were filled with advertisements for water purification devices—and the producers of this equipment reported booming sales. Marketers of bottled water benefitted too: this has become a very popular commodity in the 1990s, despite some tests that cast doubts on whether bottled water is really an improvement over tap water (Eder 1996, 8). More recent media have reported cadmium, high concentrations of chlorine, detergents, and other carcinogens, along with insect parts, dirt, and bacterial contamination in city drinking water. And this problem is not limited to Seoul. In an incident that drew international media attention, millions of people in the Pusan area were warned not to use tap water after massive amounts of phenol, a chemical known to cause cancer and damage to the central nervous system, were dumped into the Naktong River, the source of that major port city's water supply (*Los Angeles Times,* March 22, 1991). While this particular incident created unusually hazardous conditions, the contamination of municipal water throughout the country is caused by heavily polluted rivers, with industrial waste contributing some of the most toxic chemicals, and runoff from agriculture and livestock by-products adding to the effluent, too.

Agricultural Pollution and Tainted Food

The policy of the South Korean government has been to increase agricultural productivity artificially, particularly for grain (rice). Rapid urbanization and population growth led to major increases in food demand and a declining agricultural workforce. The result is capital- and technology-intensive food production. In particular, the use of fertilizers and pesticides have increased dramatically, as farmers strive to meet government demands for continuously higher yields. The heavy application of these chemicals has led to incidents in which mishandling resulted in fatal poisoning of farm workers. More significantly, these contaminants contributed to long-term, widespread rural ecological damage to ground water supplies and higher local cancer risks. The use of technology-intensive cultivation methods also results in chemical residues in food and other consumer goods (Bello and Rosenfeld 1990, 96–97; Korean Anti-Pollution Movement Association 1990, 47–54). Besides contamination from pesticides and fertilizers in rice and fruit, capital-intensive agribusiness' use of antibiotics in farm animals and various additives and preservatives may further compromise food in a variety of ways (some of which may not yet be well understood).

A SHORT HISTORY OF SOUTH KOREAN ENVIRONMENTALISM

Environmental groups have existed in South Korea since the 1970s. For the most part, however, the earliest groups were small and connected with churches or based in universities. The primary mission of these small isolated groups was to raise the ecological consciousness of the citizenry. General recognition of well-organized environmental groups, seen as more or less legitimate voices in the South Korean political scene, came during the 1980s. A loose environmental "movement" gradually coalesced during this decade, as Seoul-based organizations founded by scientists, scholars, and church leaders and initially dedicated to research and promoting public awareness connected with locally based, grassroots, antipollution activism organized by industrial workers and residents of threatened communities. After the tumultuous "democratic opening" in the late 1980s, the basic dynamic for all social movements, including environmental groups, was dramatically altered. In the 1990s the environmental movement has continued to grow and expand, attempting to broaden its appeal, routinize and consolidate its organizational structures, and operate in a new climate in which everyone, including the general population and political and corporate leaders, profess to be "proenvironment."

The Emergence of Environmental Movements in the 1980s[3]

An upswing in ecological activism and awareness swept Western Europe and the United States in the late 1960s and early 1970s (Dalton 1994), with some analysts identifying the first Earth Day in 1970 as a watershed event (Gottlieb 1993). South Korea lagged behind by about a decade. In 1980 the Study Group of Pollution (SGP) became the first environmental group to be widely recognized. It began in the 1970s as a small nonpublic gathering on the Seoul National University campus that met to discuss the social meaning of science and technology and the appropriate role of scientists and engineers in society; the members were students in natural science and engineering, whose participation in other Korean student movements was very low. While many members of the study group took a moderate view emphasizing the importance of technical expertise to solve ecological problems, others became more radicalized and committed to organizing an environmental movement (Lee interview with Byung-Ok Ahn, March 1991).

In 1982 the Korean Pollution Research Institute (KPRI) was founded. With support from Seoul-based church leadership, this was the first Korean environmental group to have staff, office space, and funding. One founding member of KPRI was Choi Yul, who is now probably the country's most visible environmental movement leader and currently the Secretary-General of the Korean Federation for Environmental Movements (KFEM)—the largest environmental-movement organization today. In its early years the KPRI operated under the authoritarian regime of Chun Doo Whan, so, despite the protective shield of church support, its activities were limited. The organization provided field support to local and regional ecological groups to help mitigate the effects of pollution and played a connecting role between these relatively isolated groups. It also promoted environmental awareness via pollution-related counseling and public lectures.

One of KPRI's most notable successes was its fieldwork at the Onsan Industrial Complex (on the southeastern coast near Ulsan). In 1985, KPRI issued a series of epidemiological reports linking the "Onsan illness," which had by 1985 stricken about 500 area residents, with industrial discharges of heavy metals such as cadmium (KPRI 1986, 86–119). This connection was confirmed by a Japanese scientist famous for his research on a similar disease in his homeland. The story soon received heavy press coverage,[4] with journalists warning of the potential danger of pollution-related illnesses throughout the country. Just as Love Canal became an icon for the U.S. antitoxics campaign (Szasz 1994), the "Onsan illness" resonated with the South Korean public and became the hottest social issue of the year. With public pressure mounting, the government's Agency of Environment was forced to respond. They did a rushed epidemiological survey of the Onsan residents that purported to show "normal" levels of heavy metal in their blood and urine. In another parallel to U.S. cases (Szasz 1994), this deceptive attempt to "whitewash" the problem failed. The public saw through it and trust in the state's ability to protect the citizenry from ecological hazard declined further. The government eventually agreed to resettle 40,000 Onsan residents in an operation portrayed as an "exodus from pollution" (Cho 1990, 209).

In 1984 various small ecological groups organized by white-collar workers and university students (most notably the SNU-based SGP) formed the Korea Anti-Pollution Movement Council (KAPMC), hoping to create an "umbrella organization" to coordinate activities. However, the "hidden" split in the SGP between moderates and radicals became obvious and soon led to conflict. The more radical faction founded a new organization in 1987 which eventually became the Korean Anti-Pollution Movement Association (KAPMA) a year later. In the late 1980s this group was the largest mass-based Korean environmental organization, with the most diverse antipollution and antinuclear activities. Its membership was more than 1300 in 1991 and it published a national newsletter ("Survival and Peace"). The more moderate SGP members were absorbed into the Korean Environment and Pollution Studies Association, which focused on research and technical/theoretical issues. KPRI continued to exist throughout this period, lengthening its name in 1989 to the Korean Anti-Nuclear and Anti-Pollution Peace Movement Research Institution. This group's continued religious ties (which had been so politically useful during the Chun years) now probably limited its mass appeal.

Though all these groups formed in the 1980s were nominally "national" ones, in reality each was Seoul-based and drew membership and support from a middle-class professional and intellectual constituency. During this period (and in some cases quite a bit earlier!), other equally important types of new environmental activism were developing in various locales across South Korea.

The origins of these emerging environmental organizations can be found in the antipollution struggles waged by local residents who were the direct victims of the worst environmental cataclysms (KPRI 1986, 263). To use the terminology of Schnaiberg and others, these are the "citizen-worker" movements mobilized to protect "use-value." In Korea, seemingly isolated peasant and fishing communities near major industrial complexes (Ulsan, Pusan, Yeocheon, etc.), motivated by a passionate desire to conserve their traditional livelihood, became fonts of local antipollution struggles. Even though some of these seemed to be very narrowly-focused, place-centered struggles (with parallels to "NIMBYism" in Western societies), and they lacked the continuity or organization to clearly qualify as "movements," these rural activists were "pioneers" of an important strand of Korean environmentalism. In the 1970s, when raising any question about government-backed development invited state repression, these groups battled against enormous odds. These early village antipollution mobilizations created strategies of active local resistance used by later generations to resist the siting of nuclear facilities and golf courses in their communities.

Sometimes these local campaigns developed into more enduring regional environmental organizations, too. For example, the successful 1983 mobilization to stop the government from approving the construction of a Jinro Alcohol Plant near Youngsan Lake, the reservoir for tap water for the city of Mokpo, led to the formation of a conservation group called the Association to Preserve Youngsan Lake. In 1987 this organization became the Mokpo Green Movement Council. Similar stories can be told for other provincial cities: by 1989 local activists had created municipal/regional environmental "councils" in Pusan, Kwangju, Youngkwang, Uljin, and Ulsan.

Despite this plethora of local and regional organizations, many of these groups remained independent and isolated. Though there usually was some type of linkage to or support from Seoul-based organizations, nationwide coordination was loose. The somewhat narrow vision of these regional groups was reinforced in their terminology: they often called themselves "anti-pollution" organizations—only later would the rhetoric change to the protection of nature and the environment in a more holistic way.[5] In part this may have reflected their everyday reality: these groups were designed to resist, oppose, and struggle. They tended to draw a clear line between polluters and victims. And they found the primary cause of pollution in rapidly developing Korean capitalism.

Green Expansion, Consolidation, and Institutionalization: The 1990s

The final years of the 1980s represent a crucial turning point for the South Korean environmental movement and its organizations and activists. Perhaps the most important factor was dramatic political upheaval in June 1987 that led to the end of the Chun regime, the beginning of limited democratization, and the activation of civil society (Lee 1993). For purposes of this discussion, the participation of urban professionals, bureaucrats, and managers in this "great uprising" is especially significant. Following this watershed event, these new middle strata have continued to play a central role in leading diverse social movements in the 1990s, including the environmental movement. Recruitment of large numbers of the urban white-collar class swelled the ranks and resources of environmental groups and moderated their ideological centers of gravity.

A second crucial sequence of events was the revelation of widespread tap-water contamination beginning in the summer of 1989. These incidents brought home the implications of the environmental crisis: up until this time, for most of the public pollution was someone else's political issue—suddenly it became each citizen's personal problem. New reports of municipal water problems in 1990, and the Nakdong River phenol incident in 1991, heightened awareness, increased support for environmentalists, and led to increased skepticism about the government's ability to protect the environment.

Just as TV in the United States helped make toxic waste a major issue a few years earlier (Szasz 1994), increasing environmental sensitivity in the South Korean media in the 1980s, focusing on lifestyle threats like unsafe drinking water, played an important role in boosting ecological consciousness. Ku (1994) did a content analysis of major daily South Korean newspaper coverage of environmental issues (Table 12.1):

Table 12.1

Trends in Newspaper Coverage of Environmental Issues

Year	1982	1983	1984	1985	1986	1987	1988	1989	1990	1991	1992
No.	479	406	369	299	433	873	1313	3250	5331	6464	8884
% ann. increase		-15%	-9%	-19%	45%	102%	50%	148%	64%	21%	37%

These data show regular increases beginning in 1986; compared with the previous year, coverage of environmental problems doubled in 1987. There is a tenfold increase in the number of articles between 1987 and 1992. Newspaper editorials show a similar trend: there were 42 in 1987 and big increases over the next two years. In 1991 Ku counted 101 editorials addressing ecological issues. While the "play" that environmental problems received in the major newspapers and other mass media (like television) was probably most influential, it is also worth noting that by August 1995 more than 20 papers were devoted exclusively to environmental issues. Though these newspapers are mostly weeklies with limited circulations, they show the blossoming public concern about ecological matters in South Korea.

By the 1990s environmentalism in South Korea had been transformed. Ecological sensitivity, which had been extremely unusual just a decade before, now permeated the entire society. Environmental awareness, which ordinary people previously considered as an esoteric philosophy of intellectuals or a political mantra of radicals, had become a mundane assumption of daily life. It touched on the shared concerns of all citizens and was part of their everyday reality and discourse. Even government and business began to present a pro-environmental posture, at least in gesture and rhetoric if not always in terms of decisions and action. The broad and wide support for environmentalism in South Korea provides the opportunity for the movement to move from an oppositional force to a mainstream political player, increasingly lacking any discernable ideological or political underpinnings. Today even the words "environment" (*Hwankyung*) or "environmental protection" (*Hwankyungboho*) convey no hint of opposition or militancy.

This process of the depoliticization and ideological moderation of Korean environmentalism, and the key role that the changing membership base played, can be illustrated in the evolution of the large mass-based KAPMA. By 1991 an increasingly mainstream KAPMA, expanded to over 1,300 members, reconstituted its organizational structure to incorporate young professionals, such as professors, medical doctors, lawyers, and journalists. In the run-up to the United Nations sponsored Rio Conference in 1992, the organization decided to coordinate preparation with business groups. Corporations sponsored KAPMA activities and paid the travel expenses of the environmental NGO representatives who traveled to Brazil. As a result, a number of more radical environmentalists quit, and the group became more moderate. By 1994 the KAPMA changed its name to the Korean Federation for Environmental Movements (KFEM): with 13,000 members it had grown into a major national organization. By December 1996 there were 25,000 members. KFEM has discarded the old KAPMA hostility toward big business and the state, avoided political "radicalism," and abandoned it's narrow "anti-pollution" focus. The leadership now embraces marketing and technical expertise to recruit new members as a key to organizational success. They set up an environmental think-tank, a center for proecology legal counsel, and an information center that opened up a Worldwide Web home page (called Korean Environmental Information). Education has become a major focus as the group lobbies for environmental topics in textbooks and curriculum; they have also established ecological summer camps, various lecture series, and specialized Citizens' Environment Schools in Seoul and Kangwon Province.

Beyond general environmental education, KFEM's two key strategies involve: 1) scientific research and monitoring of pollution; and 2) mainstream political influence via electoral campaigns and lobbying. Sometimes these two go together: KFEM teams measure air and water contamination, followed by a push for national and local government agencies to clean up the problems. For instance, the group focussed on water quality in 1994 and air quality in 1995. In the later campaign, a team of scientists, activists, local green movement groups, and sympathetic civic associations conducted a nationwide survey (measuring at 12,000 locations) to produce a national map of air pollution. This type of monitoring is accompanied by KFEM lobbying efforts. In 1994 the national Ministry of Environment was pressed to tighten the regulation of river discharges and improve the management of tap-water processing plants. Coordinating with local groups, KFEM also attempts to persuade local governments and agencies to take actions. For instance, governments in Pusan and Kyungnam Province have come out against a massive industrial complex that was recently proposed near Taegu on the Nakdong River (which supplies their residents with drinking water).[6] With an eye toward expanding its political influence, KFEM vigorously solicited and supported "environmental candidates" in the 1995 elections for local offices.[7] It supported 46 candidates and 32 won, including two mayors (Lee interview with Chi Boom Lee, KFEM Deputy General-Secretary, November 20, 1995). Presumably, these victories will lead to more green-oriented local governments.

While the dominant trend in KFEM is toward a more moderate mainstream ideological position and a preference for electoral and "insider" politics, it has maintained its links to some local "citizen-worker" movements opposed to particular forms of development. These tend to be linked to two important themes. First, KFEM has positioned itself as a leading voice against the ecological destruction promulgated by golf course development. This became a pressing problem during the final years of President Roh Tae Woo's presidency, as dozens of new courses were under construction, especially in rural areas near Seoul.[8] Local communities saw golf course construction as extremely disruptive (destroying forests and other habitats, creating erosion and mudslides, leading to pesticide pollution of water, traffic congestion, etc.), and they strongly opposed them almost everywhere they have been proposed. The KFEM national office and local chapters have provided support for this opposition. But these struggles often do *not* mirror the more sedate "work within the system" style of mainstream Korean environmentalism in the 1990s. On the contrary, the tactics of "direct action" are found here. It is not uncommon for local peasant groups to break into the offices of local governments and construction companies, block roads, and become involved in violent confrontation with police (Han 1994, 239–41).

The second major issue connected to strident local conflicts involves continued antinuclear activities. South Korea's high-speed industrialization was based on electricity supplied by nuclear reactors. During the 1980s they supplied more than half of the nation's electricity (Bello and Rosenfeld 1990, 103; Hart-Landsberg

1993, 268). In 1996 eleven commercial nuclear power plants were in operation and five were under construction in a country geographically smaller than the state of Ohio. Plans call for an incredible 55 nuclear plants to be in operation by 2031!

Despite (or, perhaps, because of) the rapid construction of this nuclear power industry, the general public is skeptical of it and mistrusts the government when it comes to nuclear issues. In recent years all nuclear power plant construction has been met with vehement resistance by local residents. Major environmental organizations like KPRI and KAPMC were deeply involved in opposing nuclear development almost from their inception, and this legacy continues in the contemporary environmental movement headed by KFEM.

Nevertheless, this is another case of the real impetus for resistance coming from the ordinary people in the impacted local communities. The people defending their "turf" from the perils of nuclear power and waste look much more "radical" and confrontational than the KFEM and its moderate "professional" leadership in their offices in Seoul! During the 1980s, as reactors were sprouting up like mushrooms across the peninsula, there were a variety of rallies and protests to block this growth. More recently, attempts to build nuclear-waste repositories have generated waves of angry, violent protest. As government siting plans are announced, antinuclear activists swing into action. Although a cadre of national activists is often involved who can mobilize support by appealing to national anti-nuclear sentiment, local resistance has been fierce as well: angry crowds rally, roads are blockaded with tractors or even burning tires, government buildings and police are targets of fire bombs, and so forth. For example, the government's duplicitous attempt to build a nuclear-waste dump on Anmyondo Island (authorities had told residents that they were planning a "research complex") set off a particularly violent confrontation. Local people attacked government buildings and set them on fire, looted shops, and physically battled police (with many injuries resulting). In the face of continuing turmoil on the island, the national Minister of Science and Technology resigned (in a national televised speech) and the project was scrapped. Similar (though somewhat less violent) protests subsequently derailed plans to build nuclear-waste storage facilities at Uljin and the remote island of Kulopdo.

Arguably, the antinuclear struggle is the most serious and important environmental issue facing South Korea today. Here, local resistance from ordinary people and activists willing to stand up to "the system" continue in the forefront, while the mainstream environmental organizations interested in education and eschewing an oppositional stance can only follow behind. Can local resistance by peasants and "citizen-workers" fundamentally challenge dominant forces in an economy that has become structurally dependent on the nuclear industry? Could these protests start the local activists on the path toward a Korean variant of "radical environmental populism"? The society's dominant structural forces seem to weigh against these possibilities. However, this case illustrates that making progress on this high-stakes ecological issue often *requires* ordinary people who are willing to be actively confrontational. Generic "pro-environmental" attitudes and big environmental organizations with ample resources and huge membership lists, however good they may be for other reasons, are no substitute for this sort of "people power."

LINKAGES TO OTHER ORGANIZING/MOVEMENTS: DOMESTIC AND GLOBAL

There is little doubt that the environmental movement of the 1990s germinated in the fertile soil of the 1980s among many groups organizing for wider political change and democratization. Since some of the most influential early environmentalists started in university-based groups (like the aforementioned SGP), it is reasonable to look for early connections between environmentalism and the student movement. While there were never any systematic intentional linkages of any formal type, it is likely that some individuals were active in both movements in the 1980s. During those years of authoritarian governments, "antipollution" movements were seen as crucial components of a wider opposition movement that was working to "change the system." Pollution was seen as a by-product of business monopolies, political oppression and cronyism, and Cold War anti-communism—which were all crystallized in the military-controlled authoritarian regime. So "antipollution" activists saw a coincidence of interests with other opposition groups working toward democratization: political change was seen as the crucial step to slowing or stopping this antienvironmental juggernaut. Today we see the historical shadow of this connection: many of the current leaders of green organizations in South Korea were formerly active in student movements for democratization during the 1970s and 1980s.

However, it is also fair to say that their has been a certain delinking between the environmental movement and other oppositional forces like students or workers/unions. The recent flow of the main currents of the ecological movement is away from radicalism and confrontation. While becoming more moderate, the major groups and key spokespeople have distanced themselves from radical students and labor. The prospect of broad coalitions between such groups seems to have evaporated as a limited form of political democratization opened up new channels for pluralist representation (or is it actually cooptation?).[9] At this point, given the heavy emphasis that green groups put on ecological lifestyles and consciousness, connections between environmentalism and consumerism and/or various strands of feminism and the women's movement seem more likely. The antisystemic potential of these types of groups, at least in contemporary South Korea, appears very modest.

The pivotal organizing that occurred around the U.N. Earth Summit at Rio de Janeiro in 1992 suggests that links to worldwide environmentalism and green NGOs with global reach may be increasingly important. Clearly, the preparation for the conference and its aftermath focused enormous attention on ecological issues in South Korea. Both the government and corporations were deeply involved. Perhaps this is inevitable with a Korean economy that is deeply dependent on exports to world markets (and particularly selling to "advanced" core countries where "green-sensitive products" are on the cutting edge of commodification).

Above, we show how this wider acceptance and support from government and corporate elites led to moderation. But it is at least possible that connections to truly global environmental groups (particularly more uncompromising and confrontational ones in Western countries) *could* lead to pressure for environmental changes that might challenge the fundamental logic of semiperipheral industrial accumulation. The development of this sort of ties between Korean and global environmental groups is still in its infancy, but these relationships could develop in interesting directions and present a promising topic for future research.

CONCLUSION

In this chapter we provided a brief overview of the development of the environmental movement in South Korea. Ecological activists (at both the local grassroots level and among "the ivied walls" of Korean universities) began their work in difficult circumstances under oppressive military-dominated authoritarianism. The movement was part and parcel of the 1980s democratization struggle. In the wake of broad political changes, Korean environmentalism in the 1990s has grown in scope and attracted new members and resources. But the major Seoul-based, national green groups have also become much more moderate, routinized organizations eager to practice insider politics. At the same time, the more spontaneous "citizen-worker" resistance to ecological threats to local communities (particularly regarding golf course development and siting of nuclear facilities) continues to be a source of high-energy, though sporadic and somewhat isolated, confrontations with state and corporate power. The way connections between the mainstream national leadership and these local activists are mediated and negotiated may become an increasing source of contention (particularly concerning issues of nuclear power and waste-dump siting, since the society has developed a structural dependency on the nuclear industry that will be very difficult to change).

The moderate lifestyle-oriented environmentalism, with the assistance of sympathetic media, and the increasing acquiescence and/or support of government and corporate elites has succeeded in dramatically changing public opinion about ecological issues. But despite this great success in changing attitudes, the massive environmental degradation that accompanied the rapid industrialization of South Korea's semiperipheral development has not been fundamentally changed. For example, 1996 witnessed many high-profile incidents: illegal releases of contaminated water from Siwha Lake, reports of pollution related human illness at Yeochen petrochemical complex, major fish kills in rivers, and extremely high levels of ozone-related smog in Seoul. News of these incidents alarmed a populace that now is environmentally conscious—and who had thought that pollution in South Korea was being better managed by the democratic government.

So, despite some impressive organizational accomplishments and a leading role in shifting public opinion, the environmental movement in South Korea can only claim limited success as to actual results (i.e., a cleaner environment). Indeed, the clearest ecological victories have often come in pockets of relatively confrontational (and violent!) local protests—to which the mainstream movement remains weakly and/or ambiguously connected. Given that the Korean model of industrialization and growth has always been deeply implicated in an environmentally destructive logic, some type of systemic challenge may be necessary for real progress. At this juncture, perhaps the threat of an "ecological legitimacy crisis" hanging over the elected government may be the best hope for a healthier environment.

NOTES

1. Note that many leading international "green theorists" also explicitly reject "political economy" interpretations precisely because they are too "economistic" and bound to class analysis (for a discussion, see Smith 1994). So, for example, Rudolf Bahro (1982) advocates a type of radical environmentalism that transcends "emancipation *in* economics to emancipation *from* economics" (32–34, our emphasis). The main thrust of the "new social movements" literature, of course, is more mundane, lacking a vision of epochal social and political transformation.

2. Here Gould, Schnaiberg, and Weinberg are borrowing ideas from the urban "growth machine" logic of Logan and Molotch (1987), who highlight the contradictory nature of "use-" versus "exchange-values" in land development and city growth in the United States. Their central argument highlights the conflicting interests between urban "growth machines" promoting "development" and "growth" to increase market values and build the local tax base and residents who value neighborhoods and communities as places to live. Indeed, Logan and Molotch (1987, chap. 6) claim that environmental concerns are frequently the "glue" that holds together citizen coalitions opposing corporate-backed growth politics.

3. The following discussion is based on Su-Hoon Lee's interviews with environmental activists and his reading of pamphlets, leaflets, newsletters, and other information published by environmental groups during this period.

4. *Hankook Ilbo* began to run articles on this January 18, 1985, and the story was soon picked up by every major newspaper, including conservative dailies like *Chosun Ilbo*.

5. Here is another interesting similarity to the United States: in the late 1960s/early 1970s early grassroots citizen-action groups in the United States adopted very similar titles (i.e., Campaign Against Pollution, Group Against Smog and Pollution) (Gottlieb 1993, 126.)

6. The ongoing battle over this Wicheon project ironically illustrates the limits on the degree to which environmental struggles can be depoliticized. The city of Taegu and Kyungbuk Province are campaigning for this project because they believe it will revive a declining regional economy, and they are receiving support from the national government (which environmental opponents of the proposed industrial complex argue is a blatant attempt by Kim Young Sam's government to garner votes from the populous Taegu/Kyungbuk region). A heated debate rages, with Pusan media leading the charge against the project. However this issue is resolved, it is bound to exacerbate growing regional tensions, which are already becoming a serious political problem (cf. Chon 1992).

7. This voting was another milestone of Korean political history. Local officials had always been appointed by the national president between 1961 and 1995.

8. This development was so emblematic of the times that critics began to sardonically refer to President Roh's "Sixth Republic" as the "Golf Republic"!

9. This trend in the South Korean environmental movement as a whole is in sharp contrast to the hopeful images of U.S. grassroots activists building coalitions and becoming

a potential force for political revitalization provided by both Gottlieb (1993) and Szasz (1994). Perhaps our Seoul-centric focus on national mainstream organizations leads us to miss some of the dynamism of local alternative groups.

REFERENCES

Amsden, Alice. 1989. *Asia's Next Giant: South Korea and Late Industrialization*. New York: Oxford University Press.

Appelbaum, Richard, and Jeffrey Henderson, eds. 1992. *States and Development in the Asian Pacific Rim*. Newbury Park, CA: Sage.

Bahro, Rudolf. 1982. *Socialism and Survival*. London: Heretic Books.

Balassa, Bela, and John Williamson. 1987. *Adjusting to Success: Balance of Payments Policy in the East Asian NICs*. Washington, DC: Institute for International Economics.

Bello, Walden, and Stephanie Rosenfeld. 1990. *Dragons in Distress: Asia's Miracle Economies in Crisis*. San Francisco: Institute for Food and Development Policy.

Cho, Hong-Sup. 1990. A Debate on Nuclear Energy: Technological Orientation or Ecological Orientation? *Society and Thoughts*: 183–91 (in Korean).

Choi, Byung Doo. 1992. Rethinking the Korean Environmental Movements: Background, Evolution, Strategies. *Shilchonmunhak (Praxis and Literature)* (Winter): 301–31 (in Korean).

Choi, Byung Doo. 1993. Philosophical Foundations and Prospects of Environmental Movements: A Reassessment of Marxism and New Social Movements. *Yiron* (*Theory*) 6: 235–56 (in Korean).

Chon, Soohyun. 1992. Political Economy of Regional Development in Korea. In *States and Development in the Asian Pacific Rim,* ed. Richard Appelbaum and Jeffrey Henderson, 150–75. Newbury Park, CA: Sage.

Cumings, Bruce. 1984. The Origins and Development of Northeast Asian Political Economy: Industrial Sectors, Product Cycles, and Political Consequences. *International Organization* 38: 1–40.

Cumings, Bruce. 1989. The Abortive Apertura: South Korea in Light of the Latin American Experience. *New Left Review* 173: 5–32.

Dalton, Russell. 1994. *The Green Rainbow: Environmental Groups in Western Europe*. New Haven, CT: Yale University Press.

Deyo, Frederic. 1989. *Beneath the Miracle: Labor Subordination in the New Asian Industrialism*. Berkeley: University of California Press.

Diani, Mario. 1995. *Green Networks: A Structural Analysis of the Italian Environmental Movement*. Edinburgh: Edinburgh University Press.

Eckert, Carter. 1992. Korea's Economic Development in Historical Perspective, 1945–1990. In *Pacific Century,* ed. M. Borthwick, 289–308. Boulder, CO: Westview.

Eder, Klaus. 1993. *The New Politics of Class: Social Movements and Cultural Dynamics in Advanced Societies*. Newbury Park, CA: Sage.

Eder, Norman. 1996. *Poisoned Prosperity: Development, Modernization, and the Environment in South Korea*. Armonk, NY: M.E. Sharpe.

Evans, Peter. 1987. Class, State and Dependence in East Asia: Lessons for Latin Americanists. In *The Political Economy of the New Asian Industrialism,* ed. F. Deyo, 203–27. Ithaca, NY: Cornell University Press.

Evans, Peter. 1995. *Embedded Autonomy: States and Industrial Transformation*. Princeton, NJ: Princeton University Press.

Foster, John. 1994. *The Vulnerable Planet: A Short Economic History of the Environment*. New York: Monthly Review.

Gereffi, Gary, and Donald Wyman, eds. 1990. *Manufacturing Miracles: Paths of Industrialization in Latin America and East Asia*. Princeton, NJ: Princeton University Press.

Gereffi, Gary, and Miguel Korzeniewicz, eds. 1994. *Commodity Chains and Global Capitalism*. Westport, CT: Greenwood Press.

Gottlieb, Robert. 1993. *Forcing the Spring: The Transformation of the American Environmental Movement*. Washington, DC: Island Press.

Gould, Kenneth, Allan Schnaiberg, and Adam Weinberg. 1996. *Local Environmental Struggles: Citizen Activism in the Treadmill of Production*. New York: Cambridge University Press.

Han, Do Hyun. 1994. Environmental Movements Against Golf Course Development Since the Late 1980s in Korea. In *Environment and Development,* 231–46. Seoul: Korean Sociological Association.

Hart-Landsberg, Martin. 1988. South Korea: The Miracle Rejected. *Critical Sociology* 15(3): 29–51.

Hart-Landsberg, Martin. 1993. *The Rush to Development: Economic Change and Political Struggle in South Korea*. New York: Monthly Review Press.

Kim, Eun Mee. 1997. *Big Business, Strong State: Collusion and Conflict in South Korean Development, 1960–1990*. Albany: State University of New York Press.

Kim, Keun Bae. 1991. Current Status and Future Prospects for Environmental Movements in Korea. *Kyongjewasahoi* (*Economy and Society*) 12 (in Korean).

Kitschelt, Herbert, and Staf Hellemans. 1990. *Beyond the European Left: Ideology and Political Action in the Belgian Ecology Parties*. Durham, NC: Duke University Press.

Korea Anti-Pollution Movement Association (KAPMA). 1990. *Compendium in Commemoration of 1990 Earth Day: For this Earth, for this Sky, and for All of Us*. Seoul: KAPMA (in Korean).

Korea Federation for Environmental Movements (KFEM). 1993–1997. *Hwankyungundong* (*Environmental Movement*). Monthly newsletter. Seoul: KFEM (in Korean).

Korea Pollution Research Institute (KPRI). 1986. *Pollution Map of South Korea*. Seoul: Ilwolseogak (in Korean).

Ku, Do-Wan. 1994. The History and Characteristics of Environmental Movements in Korea. PhD dissertation, Department of Sociology, Seoul National University (in Korean).

Lee, See-Jae. 1992. Social Change in the 1990s and Tasks and Prospects for Social Movements. In *State and Civil Society in Korea,* 441–66. Seoul: Hanul (in Korean).

Lee, Su-Hoon. 1993. Transitional Politics of Korea, 1987–1992: Activation of Civil Society. *Pacific Affairs* 66(3): 315–67.

Lim, Hyun-Chin. 1985. *Dependent Development in Korea: 1963–1979*. Seoul, Korea: Seoul National University.

Logan, John, and Harvey Molotch. 1987. *Urban Fortunes*. Berkeley: University of California Press.

Mies, M. 1986. *Patriarchy and Accumulation on a World Scale: Women in the International Division of Labor*. London: Zed Books.

Mouzelis, Nicos. 1986. *Politics in the Semi-Periphery*. London: Macmillan.

Offe, Claus. 1985. New Social Movements: Changing Boundaries of the Political. *Social Research* 52: 817–68.

Polanyi, Karl. [1944] 1957. *The Great Transformation*. Boston: Beacon Press.

Schnaiberg, Allan, and Kenneth Gould. 1994. *Environment in Society: The Enduring Conflict*. New York: St. Martin's Press.

Schnaiberg, Allan. 1980. *The Environment: From Surplus to Scarcity*. New York: Oxford University Press.

Shiva, Vandana. 1988. *Staying Alive: Women, Ecology, and Development*. London: Zed Books.

Smith, David A. 1994. Uneven Development and the Environment: Toward a World-System Perspective. *Humboldt Journal of Social Relations* 20(1): 151–75.

Smith, David A., and Su-Hoon Lee. 1990. Limits on a Semiperipheral Success Story? State Dependent Development and the Prospects for South Korean Democratization. In *Semiperipheral States in the World-Economy,* ed. William Martin, 79–95. Westport, CT: Greenwood Press.

Smith, David A., and Su-Hoon Lee. 1991a. Moving Toward Democracy? South Korean Political Change in the 1980s. *Comparative Urban and Community Research* 3: 164–87.

Smith, David A., and Su-Hoon Lee. 1991b. Antisystemic Movements in South Korea: The Rise of Environmental Activism. Paper presented at the PEWS XV Conference, University of Hawaii, March 1991.

So, Alvin, and Stephen Chui. 1995. *East Asia and the World Economy*. Thousand Oaks, CA: Sage.

Szasz, Andrew. 1994. *Ecopopulism: Toxic Waste and the Movement for Environmental Justice*. Minneapolis: University of Minnesota Press.

Wade, Robert. 1990. *Governing the Market: Economic Theory and the Role of Government in East Asian Industrialization*. Princeton, NJ: Princeton University Press.

Wallerstein, Immanuel. 1979. *The Capitalist World Economy*. Cambridge, UK: Cambridge University Press.

World Bank. 1982. *World Development Report*. Washington, DC: World Bank.

Index

About the Contributors

Albert J. Bergesen is professor of sociology at the University of Arizona and the author, with M. Herman, of "Immigration, Race, and Riot: The 1992 Los Angeles Uprising," *American Sociological Review*.

Stephen G. Bunker is professor of sociology at the University of Wisconsin. He is the author of *Underdeveloping the Amazon* and, more recently, of *Peasants Against the State*.

Sing C. Chew is professor of sociology at Humboldt State University and editor of the *Humboldt Journal of Social Relations*. His most recent article is "Accumulation, Deforestation, and World Ecological Degradation, 1500 B.C. to A.D. 1990," *Advances in Human Ecology*.

Paul S. Ciccantell is an assistant professor of sociology at Kansas State University. He recently published a co-edited volume, *Space and Transport in the World-System* (1998), and is currently conducting comparative research on the sustainability of coal-based development in Canada and the United States.

Zsuzsa Gille is a Ph.D. candidate in sociology at the University of California, Santa Cruz. Her publications include "Cognitive Cartography in a European Wasteland" in *Global Ethnographies*, M. Burawoy (ed.), and "Two Pairs of Women's Boots for a Hectare of Land: Nature and the Construction of the Environmental Problem in State Socialism," *Capitalism, Nature, Socialism.*

Walter L. Goldfrank is professor of sociology at the University of California, Santa Cruz and the author of several recent articles on Chilean export agriculture.

David Goodman is professor of environmental studies at the University of California, Santa Cruz and the recent author, with M. Redclift, of *Re-Fashioning Nature: Food, Ecology, Culture.*

Peter E. Grimes received his Ph.D. in sociology from Johns Hopkins University. In 1992 he received a grant from the National Science Foundation to apply world-systems analysis to issues of global warning. He published (with C. Chase-Dunn) "World-Systems Analysis" in the *1995 Annual Review of Sociology,* and (with J. T. Roberts) "Carbon Intensity and Economic Development 1962–1991" in *World Development.*

Su-Hoon Lee is professor of sociology at Kyungnam University and Director of the university's Institute for Far Eastern Studies in Seoul, Korea. His recent publications include "For a Humane World-System" and "Crisis in Korea and IMF Control."

Ilmo Massa is a research fellow and assistant professor of social policy at the University of Helsinki. He specializes in economic and environmental sociology.

Gavan McCormack is professor of Japanese history at the Research School of Pacific and Asian Studies, Australian National University. In 1996 he published *The Emptiness of Japanese Affluence,* which is also translated into Japanese, Korean, and Chinese.

Laura Parisi is a doctoral candidate in the Department of Political Science at the University of Arizona and a visiting instructor in the Departments of Political Science and Women's Studies at Hollins University. She recently published (with A. Gerlak) "An Umbrella of International Environmental Policy: The Global Environment Facility at Work" in *Environmental Policy and Administration in Three Worlds: Developing, Industrial, and Postindustrial,* D. Soden and B. Steel (eds.), and "Are Women Human? It's Not an Academic Question" in *Human Rights Fifty Years On: A Radical Reappraisal,* T. Evans (ed.).

Sven Peek is the Provincial Coordinator for KwaZulu/Natal of the Environmental Justice Networking Forum. During 1994–96 he was the principal field researcher for the Institute for Social and Economic Research of the University of Durban-Westville in its study of environment and development in South Durban for the City of Durban study for Local Agenda 21.

J. Timmons Roberts is associate professor of sociology and Latin American studies at Tulane University. With Amy Hite, he co-edited the forthcoming anthology, *From Modernization to Globalization: Social Perspectives on International Development.*

Christine Root has been Associate Director of the Washington Office on Africa (1972–81). During 1994–95 she was a research fellow at the Institute for Social and Economic Research of the University of Durban-Westville, where she studied community environmental participation in Durban.

Robert K. Schaeffer is associate professor of global sociology and sociology of the environment at San Jose State University and author of *Understanding Globalization: The Social Consequences of Political, Economic, and Environmental Change* and *Power to the People: Democratization Around the World.*

David A. Smith is associate professor of sociology and urban planning at the University of California, Irvine. He is currently involved in research on industrial upgrading in East Asia, global trade and commodity chains, and comparative urbanization.

Andrew Szasz is associate professor of sociology and provost of College Eight at the University of California, Santa Cruz. His book, *EcoPopulism: Toxic*

Waste and the Movement for Environmental Justice, won the Association for Humanistic Sociology's book award for 1994-95.

Immanuel Wallerstein is Director of the Fernand Braudel Center, Binghamton University, and author, most recently, of *Utopistics, or Historical Choices of the Twenty-First Century.*

David Wiley is professor of sociology and director of the African Studies Center at Michigan State University. He was a research fellow at the Institute for Social and Economic Research of the University of Durban-Westville in 1994–95 while conducting research on environmental struggles in Durban. He edited *Southern Africa: Society, Economy, and Liberation* and co-authored *Academic Analysis and U.S. Foreign Policy-Making on Africa.*

Recent Titles in
Contributions in Economics and Economic History

The Role of Precious Metals in European Economic Development: From Roman Times to the Eve of the Industrial Revolution
S. M. H. Bozorgnia

Energy and Organization: Growth and Distribution Reexamined
Bernard C. Beaudreau

Silver and Gold: The Political Economy of International Monetary Conferences, 1867–1892
Steven P. Reti

Inequality and Equity: Economics of Greed, Politics of Envy, Ethics of Equality
Charles T. Stewart, Jr.

Democracy in Desperation: The Depression of 1893
Douglas Steeples and David O. Whitten

The Origins of American Public Finance: Debates over Money, Debt, and Taxes in the Constitutional Era, 1776–1836
Donald R. Stabile

The Pillage of Sustainability in Eritrea, 1600s–1990s: Rural Communities and the Creeping Shadows of Hegemony
Niaz Murtaza

Opening the West: Federal Internal Improvements Before 1860
Laurence J. Malone

Oil and Coffee: Latin American Merchant Shipping from the Imperial Era to the 1950s
René De La Pedraja

The American Peasantry
Southern Agricultural Labor and Its Legacy, 1850–1995: A Study in Political Economy
Ronald E. Seavoy

Keynes: A Critical Life
David Felix

Treasure from the Painted Hills: A History of Calico, California, 1882–1907
Douglas Steeples

ISBN 0-313-30725-3
90000>
EAN
9 780313 307256
HARDCOVER BAR CODE